The hidden world of
mosses

ISBN: 978-1-910877-45-6

© Royal Botanic Garden Edinburgh, 2023
Published by the Royal Botanic Garden Edinburgh
20A Inverleith Row, Edinburgh, EH3 5LR
rbge.org.uk

Proceeds from sales of this book will be used to support the work of the Royal Botanic Garden Edinburgh.

The Royal Botanic Garden Edinburgh is a Non Departmental Public Body (NDPB) sponsored and supported through Grant-in-Aid by the Scottish Government's Environment and Forestry Directorate (ENFOR).

The Royal Botanic Garden Edinburgh is a Charity registered in Scotland (number SC007983).

EDITED BY
Sarah Worrall, Royal Botanic Garden Edinburgh

DESIGNED BY
Caroline Muir, Royal Botanic Garden Edinburgh

PRINTED BY
McAllister Litho Glasgow Limited

Royal
Botanic Garden
Edinburgh

The hidden world of
mosses

Neil Bell

with photography by
Des Callaghan

The moss *Pseudomalia webbiana* is restricted to the Azores, Madeira, the Canary Islands and Georgia in the Caucasus. It's relatively unusual for a moss in having strongly flattened stems with rounded leaves, and might be mistaken for a liverwort. Image: © Des Callaghan

Contents

INTRODUCTION   7

WHAT MOSSES ARE   11
Different Ways of Seeing   12
Form and Ecology   16
Lifecycle   25
Mosses, Liverworts and Hornworts   34
What Mosses *aren't*   42
Evolution   44
Groups of Mosses and Liverworts and
How they are Related   50

WHY MOSSES MATTER   81
Mosses and Us   82
Denizens of the Moss Forest   90
Other Interesting Animal–Moss Interactions   102

MOSS HABITATS   111
From Rainforests to Tundra   114
Tropical Cloud Forest   118
Wetlands   139
Temperate Rainforest   152
Mountains and Tundra   176
Disturbed and Urban Environments   195

FINAL THOUGHTS   208

REFERENCES   210

INDEX   214
Index to scientific names of mosses
and liverworts   228

ACKNOWLEDGEMENTS   232

Introduction

This is a book about something that everybody has heard of, and most people can recognise, but very few people know about.

For some, 'moss' is something that grows on the roof and occasionally needs to be scraped out of the gutters. If you are a gardener, you may have used *Sphagnum* moss to retain moisture in plant pots and have heard that more environmentally friendly options are now encouraged, but without really knowing why. Or maybe moss is associated with a romantic image of the deep forest – something that carpets the ground and hangs from the trees in the silent interior of the primeval wood.

All of these impressions skirt around the true nature of what mosses and their relatives are. They don't represent the fantastic *diversity* of mosses or tell their individual stories, and they can't reveal how strangely different mosses are from other green plants. Did you know that there are nearly twice as many species of mosses and their relatives on earth as there are ferns, and more than three times the number of conifers?

I started to really notice mosses 25 years ago when I was doing botanical survey work in Scotland. As well as categorising general habitat types I had to write short 'target notes', including lists of all of the prominent and interesting plant species at a particular site. Some of the remote locations in the hills of the Southern Uplands I was looking at were dominated by beautifully intricate but small and low-growing mosses, and yet it was the other, more obvious and familiar plants I was encouraged to list, even when there were clearly far fewer species of them there! The mosses were 'difficult', I was told, and it seemed that many people simply ignored them.

After that I read about mosses and tried to identify some of the species around me in Scotland. I had decided to pursue a career in academic biology, and before I knew it I was studying for a PhD at the Natural History Museum in London on the evolution of a small family of mosses that's mainly found in Australasia, southern South America and South East Asia and was seen as central to the question of how the most diverse group of mosses (the pleurocarps) first evolved. Since then, I've been extraordinarily lucky to have done fieldwork in such exotic places as Tierra del Fuego, Tasmania and New Caledonia, and to have spent nine years working at the Botanical Museum in Helsinki, Finland. I'm now based at the Royal Botanic Garden Edinburgh, focussing on the truly globally significant moss and liverwort flora we have right here in Scotland.

Most of the available books on mosses and their relatives fall into one of a limited number of categories. There are academic textbooks aimed at students, specialised ecological texts and highly technical floras, as well as a number of excellent field guides and other resources aimed at the beginner who is trying to identify the species in their home country or region. Not falling into any category is Robin Wall Kimmerer's wonderful *Gathering Moss*, a series of personal reflections from her point of view as a Native American bryophyte ecologist. In the UK we have Ron Porley and Nick Hodgetts's *Mosses and Liverworts* in the Collins New Naturalist series, providing an excellent introduction for the British amateur ecologist. What is lacking, however, is an attractive and richly illustrated book that introduces mosses and their relatives to the curious reader who isn't primarily looking to identify individual moss species but wants to learn about the overall diversity of these plants from a global and evolutionary perspective, what's different and special about them and why it's important for mosses to be protected.

If this book inspires you to want to learn more about these amazing and beautiful plants and especially how to identify them, the best way to get started is to reach out to any amateur societies or groups that exist where you are. If you are in Britain you should join the British Bryological Society, which organises several field meetings for members each year and also has a number of local groups scattered around the country. Perhaps more than for most other groups of organisms, you will learn much more quickly when you have other people to help you. The distinctions between many closely related species are quite subtle and are often best *shown* rather than described.

Neil Bell
Royal Botanic Garden Edinburgh

Schlotheimia ferruginea, growing on a boulder in forest on La Réunion. Image: © Des Callaghan

What mosses are

Petalophyllum ralfsii – not a moss, but a liverwort. Mosses, liverworts and hornworts are collectively known as bryophytes. This particular species is quite rare in the British Isles and is typically found in coastal dune slacks. Image: © Des Callaghan

Different Ways of Seeing

Mosses are perhaps the most misunderstood and misrepresented of all groups of organisms.

People tend to think of 'moss' as a substance, rather than as individual plants – something that is soft, cool and springy at best, but ugly, slimy and somehow mildly unsavoury at worst. When mosses are mentioned in literature it is often to convey a sense of dank, gloomy neglect, with words such as 'oozing' and 'creeping' predominating.

Nothing could be further from the truth of what mosses really are!

The mosses, considered together with their relatives the liverworts and the hornworts, are a group of around 20,000 species of green plants with a spectacular diversity of beautiful and surprising body plans, shapes and colours. They split off from the evolutionary lineage that led to all of the other land plants hundreds of millions of years ago, and as a result have been doing their own, rather different, thing ever since. Because of the particular ways they are adapted to use water they are nearly always what we, as humans, would consider small. Mostly ranging from a few millimetres to a few centimetres high, their scale means that we are aware of them, without making it easy for us to see them as they really are – unless we make the effort. Mosses

Philonotis rigida. Image: © Des Callaghan

live in the liminal zone between the visible and the invisible, in a parallel botanical reality we can gain entry to once we discover it. With the aid of a hand lens, a camera, or simply the act of looking a little more closely, we can travel to and explore the hidden world of the mosses. This is a strange world with its own miniature forests filled with its own grazers and predators, its own ecological norms and its own mechanics, at a size where water, the ultimate canvas of life, behaves rather differently from what we are used to.

One dilemma I had right from the start of writing this book was what to call it and how to refer to the plants that it deals with. Strictly speaking, this is a book about *bryophytes* – the group that includes the mosses, the liverworts and the hornworts. These three groups of plants are very distinct, although they all have much more in common with each other than they do with any other plants, and it now seems more likely than not that they

---

Mosses live in the zone between the visible and the invisible, in a parallel botanical reality we can gain entry to once we discover it.

evolved from a single common ancestor. However, there is no common name for bryophytes as a whole and the term is off-putting to many non-specialists. For these reasons I'll sometimes use the word 'mosses' informally when describing aspects of mosses, liverworts and hornworts together, although as a reminder that this isn't quite correct, I'll also use the phrase 'mosses and their relatives' fairly liberally. In general, it should be clear when mosses alone are being referred to, as in these cases they will be contrasted with liverworts and hornworts.

During the making of this book, responses to the picture of *Philonotis rigida* on a proposed front cover were mixed. Some said it looked 'alien' – the spore capsule being faintly reminiscent of a giant eyeball. The world of mosses is alien to most people, which makes it a revelation to discover and a joy to explore. Once you have read this, I hope that mosses will seem less alien as well as less hidden!

This is a book about *bryophytes* – the group that includes the mosses, the liverworts and the hornworts.

The moss *Pseudobryum cinclidioides* growing in a mire in Iceland. Body plans with structures looking like stems and leaves have evolved more than once in land plants, because they are one of the best ways to orientate green photosynthetic cells towards sunlight. Leaves are effectively solar panels! Image: © Des Callaghan

Form and Ecology

Practically all mosses, and most (but not all) liverworts, are green plants with 'stems' and 'leaves'. Although scientists who study mosses talk informally about stems and leaves, these structures look like those found in other plants because they have evolved independently to perform a very similar function, not because they share the same evolutionary origin.

All green plants need to intercept light for photosynthesis and exchange gases with the atmosphere (making food from water and carbon dioxide using the energy of the sun), and there are a relatively small number of shapes that a plant can adopt that are efficient at doing this. To catch as much light as possible you need to be thin and flat, as light can't easily penetrate into the middle of things. This means you can either have a body shape that is simply like a flat plate (as some liverworts and all hornworts have), or you can have lots of smaller, flat objects (leaves) attached to another structure that is able to raise them above the ground or otherwise position them (stems, sometimes with branches). The latter form, effectively multiple solar panels on a vertical pole or a horizontal cable, is the one that most plants have adopted.

The moss *Schistostega pennata* is remarkable for having an early stage in its growth form that persists alongside the mature stems to produce tiny, round cells that act like lenses to refract and reflect light in the dim places where it grows, so that the plant appears luminous. This has given it the English name 'goblin gold'.
Image: © Des Callaghan

The moss *Achrophyllum dentatum*, with translucent leaves that are only a single layer of cells thick, supported on a linear shoot. This is the basic form of most moss plants. The frilly-looking structures at the edges of the leaves in this particular species (called gemmae) break off to help disperse the population and are able to grow into new plants. Image: Des Callaghan (CC BY-SA 4.0)

MOSSES AND WATER

The structure of mosses' and leafy liverworts' leaves differs from other plants, which is in turn related to the way they obtain and transport water. Most mosses have leaves that are only a single cell layer thick – effectively a sheet of cells. This is why mosses have a translucent, often shining appearance – some light can pass right through them. The reason for this structure is that, with a few exceptions and unlike other plants, mosses don't have a fully developed vascular system. The veins in the leaf of a flowering plant are connected to similar vessels in the stems and roots which enables it to move water and food from one part of the plant to another. In mosses the water mostly has to be absorbed into plant cells directly over their outer surfaces. Where water is transported, it is largely over the outer surface of the plant, by *capillary action* in the surface ridges between cells and sometimes between tiny projections on the surfaces of the cells or hairs on the stems. Capillary action is a process that allows water to be 'pulled along' a very narrow channel, because the force of the 'stickiness' of the water molecules that holds them together is more than the force of the weight that might separate them.

This in turn means that mosses can't be very large plants, because as anything of the same shape gets bigger, its surface area *as a proportion of its volume* becomes less. Mosses need as much surface area as possible to absorb water and to exchange gases (carbon dioxide and oxygen) for photosynthesis, because they generally don't transport water and nutrients internally.

Mosses need as much surface area as possible to absorb water and to exchange gases because they generally don't transport water and nutrients internally.

This strategy is a double-edged sword – if there is a large surface area to absorb water quickly and efficiently, this will also rapidly lose water when the environment is dry. Thus, mosses need to be able to ebb and flow with the availability of moisture. Rather than *resisting* drying out, as most other plants do, they tend more to *tolerate* drying.

It may be surprising to find out that although mosses are associated with wet places and thrive when water is continuously available, they are usually much better able to *survive* drying out than other plants are. In fact, it is quite possible for many moss species to dry out completely and remain in a desiccated state for prolonged periods, only to recover and start growing again when moist conditions return. It seems that they can do this by storing a set of blueprints for recovery (in the form of mRNA transcripts, effectively copies of the information on certain critical parts of the plant's DNA) and then using these to direct special repair mechanisms when water is available again. Large collections of mosses are kept in botanical archives (herbaria), and it has been shown that these dried specimens are sometimes able to grow again after 20 years of storage!

The difference between mosses and flowering plants in terms of their relationship to water is similar to the difference between 'cold blooded' and 'warm blooded' animals and their relationship to heat. Just as a lizard needs heat to be active and to thrive but is able to allow its body to slow down and become inactive when it is colder, so a moss needs water to thrive but is able to tolerate almost complete desiccation for prolonged periods when it is required to. This, in effect, is a different strategy for living – it provides mosses with advantages over other plants in some situations, but not in others.

Green leafy gametophytes and yellow-orange sporophytes of *Polytrichastrum sphaerothecium*. This species is in the family Polytrichaceae – the name means 'many hairs' in Greek, referring to the long hairs on the cap (the calyptra) that covers the young spore capsule. Image: © Des Callaghan

Lifecycle

TWO LIFE STAGES, TWO DISTINCT TYPES OF PLANT

If you have ever looked closely at mosses on top of an urban wall, you may have seen strange, pin-like structures with little blobs on the end seemingly growing out of them. You may have speculated that they must be some sort of fruiting body, and you would have been correct in a sense – except they are a little more than that! These little blobs-on-sticks are called *sporophytes* (meaning 'spore bearing plants'), and they represent a completely different phase of the moss's lifecycle from the green, leafy part (which is called the *gametophyte*, or 'gamete bearing plant').

Although the sporophyte is attached to the gametophyte (its mother) and nutritionally dependent

Oncophorus integerrimus growing beside a stream in Iceland. The sporophytes here have yet to open to release their spores and still retain their chaffy-looking calyptras, covering the spore capsules as they develop. The calyptra is actually derived from the green, leafy gametophyte rather than being part of the sporophyte itself and is carried upwards on top of the sporophyte as it grows.
Image: © Des Callaghan

Hookeria lucens, with green, leafy gametophytes supporting red-brown sporophytes. The sporophytes develop after sexual reproduction and produce spores which are released from the ends of the cylindrical capsules. Image: © Neil Bell

on it, it is actually a distinct individual. The 'baby' sporophyte is nourished via a placenta, but it remains that way even when it is fully grown. It's just a blob-on-a-stick after all, it doesn't have leaves to make its own food! The sporophyte, as its name suggests, eventually produces spores – thousands of tiny cells resembling specks of dust that can be transported by wind over long distances. These spores are the main method by which the moss disperses (just as seeds are for flowering plants).

Spores have a specialised, multilayer outer coating that is able to resist chemical attack, drying out and UV radiation.

FULL CYCLE

Because a lifecycle is effectively a closed circle, we could theoretically start anywhere in describing it. This is perhaps more true for mosses than for many other types of organisms, because both the gametophyte and sporophyte stages start off as single cells. However, let's start with the spores!

Spores are usually single cells (occasionally they are multicellular) with a specialised, multilayer outer coating made from an extremely tough substance called sporopollenin that is able to resist chemical attack, drying out and UV radiation – it's one of the main things that allows plants to live on land and is also found in the pollen grains of more familiar plants.

Moss spores, like eggs and sperm in humans and other animals, have only a single copy of the species' DNA. In humans, all other cells have two

copies (one from each parent), and it is in this brief coming together of the human egg and sperm that the first cell with two copies is created. In mosses this stage of the lifecycle, in which there is only a single DNA copy, is vastly extended and is the most visible phase within the life of the plant.

When the spore germinates it produces green filaments that look a little like tiny algae. This is the first stage of the gametophyte, and in turn it produces buds that grow into the green, leafy shoots that most of us think of as 'moss' – the mature stage of the gametophyte. Every cell in this mature plant only has a single DNA copy, because the cells are all derived from that single spore. But the process of coming together of male and female cells has merely been delayed. The green, leafy gametophytes now produce specialised male and female cells, or gametes – eggs and sperm – in dedicated structures. Unlike in humans, however, no special cellular process is needed to ensure that these cells contain only one copy of the DNA, because they are developed from cells that are already like this! In some species the gametophytes are always either male or female, so produce only eggs or only sperm, while in others the gametophytes are both male and female and produce both eggs and sperm on the same plant.

Whether the plant has only one sex or both, the egg is fertilised ideally by a sperm cell from a different plant. This is the stage when the first cell is produced with *two* DNA copies. This cell, of course, is what grows into the next stage of the lifecycle – the sporophyte, our 'blob-on-a-stick'.

The cells in the sporophyte all have two DNA copies, just like the cells in our own bodies (and like the cells in the green, leafy parts of more familiar plants). However, inside the spore capsule, some of these cells divide to eventually produce cells with only single DNA copies again – spores of course – and the cycle is complete!

This 'two stage' lifecycle is often referred to as 'alternating generations'. Every species of moss has its own distinct gametophyte by which it can usually be recognised, for example by the particular form and habit of its leaves, or the shape of its cells. The sporophyte generation also often has its own distinct form, with species varying in the shape of the spore capsule, the length of the stalk on which it is held, and the structures associated with spore release. Two for the price of one!

VEGETATIVE REPRODUCTION

It was mentioned above that spores are the main method used by mosses for dispersal. These are ultimately derived from sexual reproduction, helping to ensure that new colonies are made of different individuals with different genetic makeups. However, many mosses and their relatives are able to reproduce in other ways that do not involve sex, resulting in new individuals that are genetically identical to the single parent. In fact, mosses are particularly good at this!

Remarkably, the cells in almost any part of a moss are able to revert to a basic, unspecialised state in the right conditions and give rise to new plants – a process called *totipotency*. In a way, it's as if any part of the plant is able to act a little like a 'seed' (although of course mosses don't have seeds and the process is non-sexual). This means that, for example, if a moss dries out and the tip of a leaf is broken off, this tiny, chaff-like fragment may be able to be transported over quite long distances and then give rise to a new moss plant if it lands in suitable conditions elsewhere.

Pieces of moss plants that can function in this way can be roughly divided into parts that are primarily designed to perform other purposes but may act to disperse the plant in certain circumstances (like the leaf tip example above), and parts whose primary function is to become detached for dispersal. Mosses and their relatives have many different kinds of the latter. *Gemmae*, for example, are small, usually multicellular objects that may be disk-shaped, club-shaped, hair-like or globular and are produced on stems, leaf tips or often on specialised structures. Perhaps the most familiar and easy to see are the 'gemmae cups' produced by the thalloid liverwort *Marchantia*, a common plant of disturbed soil and paths in gardens, that resemble tiny, frilly bowls of lentil-shaped green gemmae. A number of mosses with short lifespans produce *tubers* on the hairs covering the underground parts of their stems – as the name suggests, these are a little like microscopic potatoes! They are often quite brightly coloured, and rather than be dispersed, are mostly designed to persist in the soil for months or years until such time as the conditions are suitable for the moss species to grow again.

Structures for vegetative dispersal called gemmae, as found on the shoot tips of the leafy liverwort *Anastrophyllum hellerianum* (left), and the leaf surfaces of the moss *Oedipodium griffithianum* (right). Images: Des Callaghan (CC BY-SA 4.0)

The importance of vegetative reproduction in mosses and their relatives is dramatically demonstrated by the existence of some species that have never been known to produce gametes (eggs and sperm) or sporophytes. These species are required to persist and disperse entirely vegetatively, which means that all new plants formed will be genetically identical. Sometimes it has been shown that such species are indeed genetically identical over large areas, while other times this is not the case, meaning that sexual reproduction must happen very rarely or must have happened in the past.

Leafy liverworts have leaves in two main ranks (often there is a third rank underneath, although these leaves are usually smaller), giving plants such as this *Gongylanthus ericetorum* a flattened, symmetrical look. Image: © Des Callaghan

Mosses, Liverworts and Hornworts

Up until now, this book has referred to mosses 'and their relatives' and only mentioned *liverworts* and *hornworts* in passing. In fact, the general term 'mosses' has sometimes been used to refer to mosses in the strict sense as well as to liverworts and hornworts by extension. Let's meet the relatives!

Mosses are the largest of the three groups in terms of species, and nearly everything we have said so far about mosses also applies to liverworts and hornworts – the three groups share the same lifecycle and general ecology, as well as many of the same structures. The term *bryophytes* is used to refer to mosses, liverworts and hornworts together. This, perhaps, is a word that is unfamiliar to most people, which is why we've occasionally used 'mosses' in a vague sense as a stand-in. As will be seen later, this isn't quite as misleading as it might have been a few years ago, because it is now known that these three types of plants are likely to be each other's closest relatives, in which case they can truly be thought of as one big group in an evolutionary sense.

DIFFERENCES

All mosses (more or less) have stems and leaves. Compared to the other two groups they are perhaps the best adapted to life on land and are found in almost every part of the world, from lush forests to rocky mountain tops, and from city centres in the tropics to Antarctic tundra. This is reflected in their diversity, with about 12,000 species known – nearly twice as many as the roughly 7,000 known species of liverworts. The hornworts are by far the least numerous of the three groups, with only about 200 species known worldwide.

Most species of liverworts also have stems and leaves, but some groups have a completely different growth form that we refer to as *thalloid*. You may have seen flat, green, rubbery-looking plants

Breutelia gnaphalea, like most mosses, has spirally arranged leaves (although some species have flattened shoots that may give the impression of leaves being in two ranks, while a few actually have leaves in two or three ranks). Image: © Des Callaghan

a few millimetres thick growing pressed to disturbed soil in your garden, or perhaps on the lower surfaces of vertical rock faces in wet places. These are likely to be thalloid liverworts. They are plants that have adopted a different strategy for exposing as much flat surface as possible to the environment – they simply spread as a thin layer over soil or rock. They can exploit these surfaces more ably and more quickly than other plants because they haven't got (and don't need) roots. Although only around 1,000 of the 7,000 species of liverworts are thalloid, some of them are quite noticeable because they are large and relatively common. But all of the other 6,000-odd liverworts have stems and leaves and look much more like mosses.

So, if 'leafy' liverworts look like mosses rather than thalloid liverworts, how do we know they are liverworts? Well, if they have *sporophytes* – that different phase of the lifecycle described earlier that all mosses and their relatives have – we can easily tell them apart. In mosses (see images on pp. 24–29), these 'blobs-on-sticks' are fairly long-lived (months), usually green when young and brownish when mature, and the 'blobs' usually more or less elongated with a little permanent opening at the end from which the spores are gradually released. In liverworts, the sporophytes are short-lived (days), and the 'blobs' look like little black spheres on the end of very delicate, transparent 'sticks' that look for all the world like tiny Chinese glass noodles. And rather than releasing the spores gradually through an opening, the spherical blobs of liverworts will suddenly rupture and release the spores in a rapidly expanding mass. This type of sporophyte (more or less) is found in all liverworts, whether or not the other part of the plant is thalloid or leafy.

But what if the tiny, translucent green plant with stems and leaves that you are looking at doesn't have any sporophytes on it? After all, that's likely to be the case for liverworts most of the time, because the sporophytes are so

The tropical thalloid liverwort *Cyathodium cavernarum*.
Image: © Des Callaghan

Megaceros gracilis, a hornwort, with young 'horns' (sporophytes) starting to grow.
Image: © Des Callaghan

short-lived. In that case there are other ways we can tell mosses and liverworts apart, although it takes some practice. In general, the leaves of mosses are arranged spirally on stems rather than in distinct ranks, while in liverworts the leaves are always in two or three ranks. This means that liverworts usually look 'flattened', with distinct upper and lower sides (although there are a few mosses that look like this too).

And what about hornworts? OK, there are only 200 species worldwide (with four in the UK) and you are unlikely to come across them very often, but that doesn't mean it's OK to ignore them! To cut a very long story short, hornworts look rather like thalloid liverworts (they never have stems and leaves), but their sporophytes are very different. Rather than 'blobs-on-sticks' they are just long green 'horns', with the spores residing in the whole length of the horn rather than in a blob at the top. When it's mature, this will split open gradually from the top to release the spores.

What Mosses *aren't*

There are some other living things that look a bit like mosses and are often found in the same places because they share a similar lifestyle. And then there are things that are commonly called 'mosses' that most certainly aren't!

Mosses aren't *algae*. The term algae encompasses many photosynthetic organisms that aren't plants. Although algae are particularly associated with aquatic environments, some species of *filamentous* (thread-like) *green algae* are common in places where mosses are also found, such as on wet rocks and bark. From a distance they could be mistaken for mosses, although they lack stems and leaves. And we have already seen that the very youngest stages of moss gametophytes closely resemble green algae.

Mosses aren't *lichens*. Although lichens live in a similar way to mosses and have a similar relationship with water, they are extremely distantly related. Lichens are a type of fungus (like mushrooms and their relatives) that have a special relationship

Mosses aren't algae. Mosses aren't lichens. Mosses aren't Spanish moss.

with algae, while mosses and their relatives are true plants. In fact (incredible as it sounds), lichens are more closely related to animals, including us, than they are to plants! Nonetheless, mosses and lichens have independently discovered a similar way of living (they have a similar *ecology*), so they often grow together and can superficially resemble each other. Mosses and lichens are often most diverse in the same places, for example growing on twigs and branches in humid forest, or directly on rocks in open habitats, and for this reason they have often been studied together historically.

Mosses and lichens often grow together and can superficially resemble each other.

Mosses aren't *clubmosses*. Despite the name, and despite looking a little like robust mosses, clubmosses are more closely related to other plants (ferns, flowering plants, etc.) than they are to mosses and their relatives.

Mosses aren't *Spanish moss*. Spanish moss is a type of flowering plant that grows on trees, and in fact looks more like some types of lichen (which it isn't either!) than moss.

Evolution

ANCIENT PLANTS?

In popular science articles, and even in botany textbooks from just a few years ago, you will often see mosses referred to as 'ancient'. They are blithely described as 'the oldest plants on earth' and 'the earliest land plants'. In fact, the evidence for thinking of mosses and their relatives in this way has always been rather circumstantial. In recent years it has become much more so, as we have learned more about the likely relationships between mosses, liverworts, hornworts and the other more familiar plant groups.

There *are* a number of good reasons for thinking that the earliest plants on land might have resembled mosses or their relatives. Mosses are quite small and simple, and they are dependent on the immediate availability of water for reproduction and sustained growth. We know that all plants on land evolved from a type of green *alga* that today lives mostly in fresh water (basically, pond scum). These algae didn't have complex *sporophyte* generations like mosses and all other land plants do (remember the 'blobs-on-sticks' and 'alternating generations'?). But like mosses and their relatives (and unlike the other land plants) they did have

> You will often see mosses referred to as 'ancient'. They are blithely described as 'the oldest plants on earth'...

complex *gametophyte* generations. So, it might appear to make sense if algae with complex gametophytes had evolved into mosses with both complex gametophytes and complex sporophytes. Plants similar to mosses might then have evolved into the more familiar land plants that have considerably more complex sporophytes and very simple, vestigial gametophytes. However, it's also quite feasible that mosses and their relatives could be descended from ancient plants that were a little more like familiar modern plants – in other words, that they have actually evolved to become smaller and more simple.

Of course, it would all be much easier if there were lots of unambiguous fossils from the period during which we think land plants were evolving (perhaps around 450–500 million years ago), but there aren't. The fossils we do have from this period are mostly just spores (they had a protective coat that protected them from decomposition long enough to become fossilised). These spores do, in fact, resemble the spores found in some liverworts. However, this doesn't necessarily mean that the plants that produced them were liverworts. It's quite possible that plants of many different types once produced spores like these and that it's only some liverworts that still do today.

...the evidence for thinking of mosses and their relatives in this way has always been rather circumstantial.

EVIDENCE FROM DNA

Until recently, one of the main reasons for thinking that the first plants on land were like mosses, liverworts and hornworts was that we believed that these three groups (the *bryophyte* groups) were *not* each other's closest relatives. Just a decade ago it seemed very likely that hornworts were most closely related to all of the more familiar plants (ferns, conifers, flowering plants, etc.), with mosses more distantly related, and liverworts more distantly related still. This would have meant that liverworts would have branched off from all other plants in the evolutionary tree before mosses branched off from the line that led to hornworts and the other plants. If you think about it, this would have made it significantly more likely that the earliest plants had the features that all of the three bryophyte groups have in common, and less likely that they had features that our familiar ferns, conifers and flowering plants all share.

Recent research using large amounts of information from DNA, however, has shown that these

> We believed that these three groups (the *bryophyte* groups) were *not* each other's closest relatives.

> This would have meant that liverworts would have branched off before all other plants in the evolutionary tree.

Evolutionary tree showing the most likely relationships of mosses and their relatives (liverworts and hornworts) to each other and to other living land plants. If this is accurate then mosses, liverworts and hornworts have all evolved from an ancestor (purple square) that wasn't shared with other land plants. But this ancestor and the other land plants in turn shared an ancestor further back in time (red square). Image: © Neil Bell

ideas were probably wrong. It now seems much more likely that mosses, liverworts and hornworts are all more closely related to each other than any of them are to any other plants that are around today – in other words, that bryophytes are a 'real thing', not merely a convenient collective term that botanists use to refer to

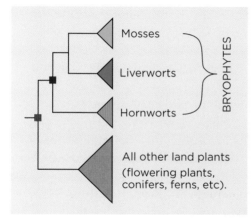

three different groups that have a lot in common. This makes most bryologists happy. More importantly, it means that hundreds of millions of years ago, primeval land plants probably split into two lineages, one of them leading to mosses and their relatives, and the other leading to ferns, clubmosses, and the familiar seed plants. In a very general sense, and without going into too much detail, this means that there's no reason to think that those primeval plants were more like modern mosses than they were like the earliest fossil plants we know of in the lineage that led to ferns and seed plants. And in fact, these fossils do seem to combine features of mosses and their relatives with those of other plants that are around today (while looking nothing much like any of them!).

A moss, *Vetiplanaxis pyrrhobryoides*, preserved in 100-million-year-old amber from Myanmar. This species belongs to a general group of mosses, the Hypnodendrales, that is still around today, although all species of *Vetiplanaxis* are now extinct. Mosses belonging to this group may have been more widespread and diverse 100 million years ago than they are today. Image: © Neil Bell and Peter York

FOSSILS

Unfortunately, the fossil record for mosses and their relatives is much sparser than for other land plant groups. This has been thought to be because of their small size and delicate nature, making it difficult for them to fossilise very well. However, some scientists disagree with this and suggest that other factors, including human ones, might be responsible (i.e. we may just not have looked hard enough!). Fossils of anything other than spores from the earliest period of bryophyte evolution are almost entirely lacking, and most fossils of things that might be mosses are dubiously identifiable as such.

Then, in rocks from about 385 million years ago onwards, we start to find things that are certainly liverworts, and shortly after this, things that are certainly mosses. Bryophytes and other plant groups were already distinct by this time, however, so arguably we have missed the most exciting part of the story! Also, the fossils are sparse and usually so different from any modern bryophyte that we can't tell all that much from them about how mosses and their relatives might have evolved.

This latter problem is particularly true for the most abundant and well-preserved group of early moss fossils, from 250–300-million-year-old rocks in Siberia. At this time, Siberia, or 'Angaraland', had only recently ceased to be a large island continent in its own right, having collided with the continents of Kazakhstania and Baltica (you can probably guess where they are now!), adding the final pieces

to the jigsaw puzzle of the great supercontinent Pangea. The moss fossils from this period are fascinating, but with a bizarre combination of features that make it impossible to say how they relate to any modern groups of mosses. And then, just after this period (which geologists call the Permian), the biggest extinction event in the history of the earth occurred (possibly thanks to huge volcanoes caused by all of those continents colliding). It's quite a long time later before we find any fossil mosses again, and they mostly look like members of modern families. Once more, geology has dozed off during one of the most exciting parts of the film!

There are, however, quite a lot of very well-preserved mosses in amber. Amber is fossilised tree resin that is able to preserve delicate structures almost completely, because they are wholly enclosed in the liquid resin before it fossilises and is effectively sealed off from damage and decay. Most moss fossils in amber are 'only' about 40 million years old and quite similar to species existing today. Nonetheless, some of the most interesting fossil mosses have been found in older (100-million-year-old) amber from Myanmar, and these may shed light on the evolution of the most successful group of mosses around today. The *pleurocarpous mosses* (see Box, p. 132) tend to have more complicated branching structures than other mosses, and some of the extinct species preserved in amber from Myanmar represent an early group of pleurocarps that was probably much more diverse then than it is now.

Groups of Mosses and Liverworts and How they are Related

Although not much can be revealed about the evolution of mosses and their relatives from fossils, we can use statistical methods to work out the most likely relationships of different groups of bryophytes to each other, based on their similarities and differences (including in their DNA), and certain assumptions about the way that evolution happens. Many of the major groups of mosses and liverworts have been recognised since the 19th century, although in recent decades we have gained a better understanding of how they relate to each other and of the finer-scale relationships within them.

Here we will outline this diversity using the broadest brush-strokes possible. In other words, we will identify the groups that are distinct from each other at the most general level – the ones produced by the earliest splits in the evolutionary tree. Some of these groups are very small and today include only a few species, while others are truly huge and include large proportions of all of the bryophytes currently existing.

Although most of the variation in mosses and their relatives is actually now within these larger groups, we can't describe all of it here. However, throughout the rest of the book, in the sections on different types of habitats, we will spotlight some of the more interesting groups of moss and liverwort species that occur within larger general groups. Think of the purpose of the current section as being similar to explaining how wolves are

Evolutionary tree showing the most likely relationships of the major groups of mosses to each other. Image: © Neil Bell

MOSSES

As outlined earlier, practically all mosses are 'leafy', with a body plan for the *gametophyte* consisting of leaves attached to shoots. Sometimes there is just a single shoot, while some species have shoots with branches that look different from the main shoots. This leafy part of the plant is the *gametophyte* of course – the other part, the *sporophyte*, consists of just a single spore capsule, usually on the end of a stalk called a *seta*. Quite a lot of the major groups of mosses are told apart from each other by the types of spore capsules they have, and

different from dogs, without listing and describing all of the different types of dogs!

especially the mechanisms they use to release the spores. In the figure below we can see the relationships of the various major groups to each other according to current ideas.

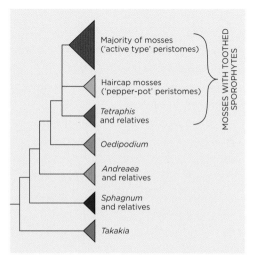

Takakia lepidozioides growing on a damp slope in British Columbia, Canada. Image: Stu Crawford (CC BY-SA 4.0)

Takakia

It seems likely that hundreds of millions of years ago, the evolutionary line leading to the two known species of *Takakia* split off from the one leading to the roughly 12,000 other species of moss that exist today. This doesn't mean that *Takakia* has remained the same for all of that time of course, and it's almost certain that the *Takakia* lineage has given rise to many more species in the past, all of which are now extinct. But it does mean that *Takakia* is more 'different', just because all other moss species share an ancestor that existed at some point after the *Takakia* lineage went its own way.

And *Takakia* really is different! So much so that it was thought to be a liverwort rather than a moss when it was discovered, and it was only when its sporophytes were found for the first time in 1993 that it was unambiguously confirmed to be a moss.

Takakia has tiny shoots with leaves that are divided into narrow filaments, and sporophytes that open to release spores by splitting along a longitudinal spiral slit, which is unique among mosses.

Spore capsules of *Sphagnum*. The lid detaches explosively, shooting the spores into the air! The green structure that looks like a seta is in fact not part of the sporophyte, but rather a pseudopodium produced by the leafy gametophyte to lift the spore capsule. The seta is absent in *Sphagnum*. Image: © Des Callaghan

Sphagnum

Sphagnum has a crucial role in creating and maintaining peat bogs, and this role will be discussed later in the book. There are about 200 species of *Sphagnum* worldwide and the lineage they are in was probably the next one to split off from all of the other mosses after *Takakia*. Like *Takakia*, *Sphagnum* has sporophytes that are unique among living mosses. They are dark and spherical, superficially resembling those of liverworts, but unlike in liverworts they release their spores by explosively popping off a little lid and firing the spores into the air. There has been some debate about whether this is achieved through compressed air in the capsule or stored elastic forces in the sporophyte walls, although recent research seems to support the second explanation. In any case, the spores can be shot an incredible 10–20 centimetres into the air, with an acceleration of 36,000G (roughly equivalent to half of the acceleration of a bullet leaving a rifle barrel)!

Sphagnum capillifolium, a fairly small species with a compact growth form that's common in bogs and wet heathland in Britain.
Image: © Des Callaghan

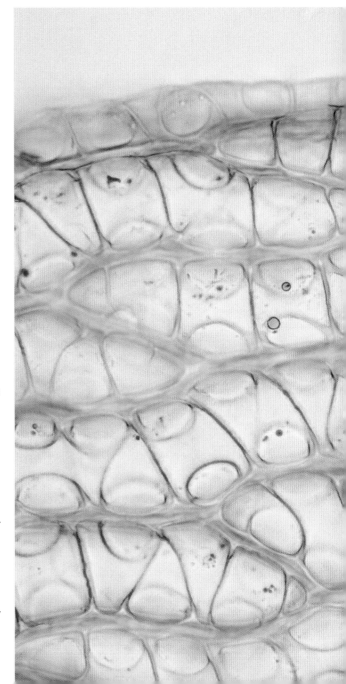

▼ The unique cell structure of *Sphagnum* leaves. The thick green lines are specialised narrow cells in which photosynthesis occurs. These surround giant, empty cells that are strengthened with ribs (the narrow lines running crossways) and fill with water when the plant is wet. Image: Hermann Schachner (CC0 1.0)

The really interesting thing about *Sphagnum*, however, is the structure of the leaves. These have two completely different types of cells in them. The 'normal' green cells that capture light and perform photosynthesis are extremely long and narrow, squashed between giant, non-green cells that are dead and empty when the plant is mature. These non-green cells are reinforced by special ribs so that they don't collapse when they are empty and have little holes in them to let water in and out. They effectively act like hundreds of little water bottles in each leaf, allowing the *Sphagnum* plant to store 20 times its own weight in water! This is a large part of the secret as to how *Sphagnum* is able to prevent organic matter in peat bogs from decaying so that it can build up to form peat. By keeping so much water on the surface of the bog, everything is constantly waterlogged, inhibiting decay. The living *Sphagnum* carpet is like a giant wet blanket over the soil.

Hundreds of little water bottles in each leaf allow the *Sphagnum* plant to store 20 times its own weight in water.

▶ Spore capsule of *Andreaea*, which opens by four slits and has been compared to a Chinese lantern. Like *Sphagnum* (p.54), *Andreaea* lacks a seta and the seta-like structure is a pseudopodium produced by the gametophyte. Image: Hermann Schachner (CC0 1.0)

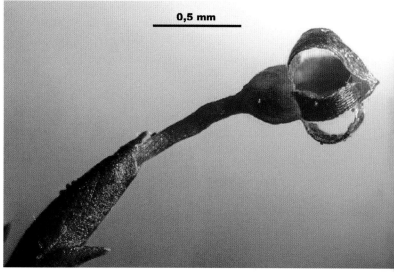

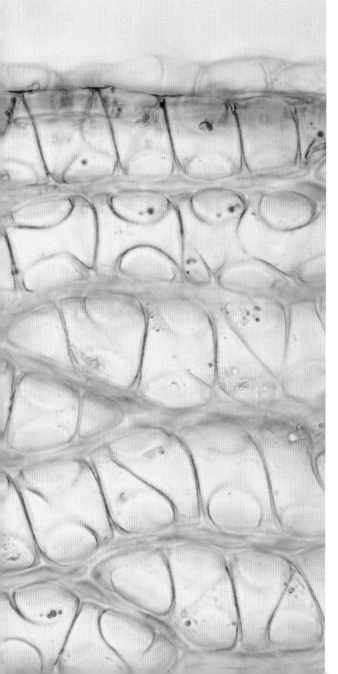

Andreaea

The roughly 100 species of *Andreaea* are rock mosses growing mostly in wet places in mountains. Their lineage is the next one to branch off from the moss evolutionary tree after *Takakia* and *Sphagnum*. While their leafy shoots are rather dull and unremarkable, like the other two outliers on the evolutionary tree they have uniquely odd sporophytes. Again, they are small and roughly spherical or oval, but they lack the detachable lid of *Sphagnum* and open instead by splitting along four vertical slits.

However, these slits don't extend to the very tip of the sporophyte, creating openings that bulge out like those in the arms of an old-fashioned doublet. Some people have thought that this makes them look like tiny Chinese lanterns, so the *Andreaea* lineage is sometimes referred to as the 'lantern mosses'.

There is another lineage, *Andreaeobryum*, that shares some features with *Andreaea* but is very different in other ways and contains only one species, *Andreaeobryum macrosporum*. Its relationships have been uncertain in the past, but it is now thought to be more closely related to *Andreaea* than to anything else.

Mosses with 'toothed' sporophytes

While the few-hundred-odd species in the groups we have mentioned so far are very interesting (and in the case of *Sphagnum* very important), the vast majority of the 12,000 species of moss alive today occur in one single, large group. These mosses all have sporophytes that open with a detachable lid, like *Sphagnum*, but they don't explosively release their spores. Instead, when the lid falls off it reveals a special, quite complicated apparatus that gradually controls the release of the spores over a prolonged period. This apparatus is called the *peristome*, and it resembles one or more rings of teeth surrounding the

Leucodon treleasei growing on a tree trunk in Madeira. The name *Leucodon* comes from the Greek 'leukos', meaning 'white', and 'odon' ('tooth'). The white-coloured peristome can be clearly seen here surrounding the small opening on the oval spore capsule.
Image: © Des Callaghan

opening at the top of the capsule from which the spores are released after the lid becomes detached.

There are two basic types of peristome. Let's call them the *pepper-pot type* and the *active type*. The pepper-pot peristome is only found in one family of around 200 species (the 'haircap mosses') and consists of many robust teeth that are all attached at their tips to a flat disk. The whole structure is static, and the spores are released from the gaps between the teeth just like a pepper-pot, when the sporophyte is shaken by the wind or passing creatures.

In the active type of peristome, the teeth are more delicate (they are actually made up of the remains of dead cells that grow specially to manufacture the material for the teeth in their cell walls). These teeth usually spread open when the air is dry, allowing the spores to escape when conditions are best for them to be blown away, and bend inwards again when it is wet, closing over the opening at the end of the sporophyte.

This type of peristome, found in the vast majority of mosses, acts efficiently to regulate the release of spores in a way that ensures that they have the best chance of travelling long distances. There are a number of different subtypes of this active type of peristome, and these subtypes characterise major groups of related species.

There is actually a third general type of peristome that is only found in one very small group – the moss *Tetraphis* and its relatives. The teeth in this group are similar in structure to the pepper-pot type, but there are only four teeth and they are not attached at their tips to a flat disk.

> Spores are released from the gaps between the teeth just like a pepper-pot, when the sporophyte is shaken by the wind or passing creatures.

Evolutionary tree showing the most likely relationships of the major groups of liverworts to each other. Image: © Neil Bell

LIVERWORTS

We've already mentioned that there are two basic forms that liverworts can have. While most species have stems and leaves like mosses, some are *thalloid* – resembling flat, rubbery-looking mats (although they don't feel rubbery!).

There are three quite distinct groups of thalloid liverworts, and they don't all form one large group that's distinct from the leafy liverworts in an evolutionary sense. Instead, some thalloid liverworts are more closely related to leafy liverworts than they are to other thalloid liverworts. The figure opposite shows the relationships of the various major groups of liverworts to each other according to current ideas.

Unlike in mosses, where there are many differences in the sporophytes between the major groups that we can use to recognise them, the sporophytes of liverworts are all rather similar, even between the leafy and thalloid groups. None open by lids or have toothy peristomes.

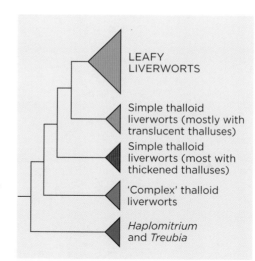

LEAFY
LIVERWORTS

Simple thalloid liverworts (mostly with translucent thalluses)

Simple thalloid liverworts (most with thickened thalluses)

'Complex' thalloid liverworts

Haplomitrium and *Treubia*

Instead, they usually open by four slits (or sometimes they rupture without slits), with the four sections of the spore capsule bending backwards and downwards. The spores are then pushed out by special cells called *elaters* – these are effectively tiny springs that are coiled up when the sporophyte is closed, and then rapidly uncoil when it opens. The spores and the elaters are all pushed out in an expanding mass by the uncoiling elaters.

▼ The evolutionarily highly isolated *Treubia*, growing in New Zealand. The yellow dots are a particularly large type of oil body, a structure found in most liverwort cells. Image: Auckland Museum (CC BY 4.0), with cropping by Neil Bell

▶ *Marchantia polymorpha* subspecies *montivagans*, a complex thalloid liverwort. As with the similar structures seen in *Conocephalum conicum* on the next page, the fingered, umbrella-like objects on stalks are not sporophytes, but parts of the gametophyte that carry the female reproductive structures (and eventually the sporophytes themselves, which are not stalked and are produced underneath the 'fingers'). Image: © Des Callaghan

Haplomitriaceae and Treubiaceae

These two families together comprise a lineage (like *Takakia* in the mosses) that split off from all of the rest of the liverworts hundreds of millions of years ago. And just as with *Takakia*, they are freakishly different from all the others! Unlike the situation with the *Takakia* lineage though, these two elements are very different from each other, as well as (considered together) different from all other liverworts. Between them they contain about 20 species, nearly all of which are only found in the southern hemisphere.

Haplomitrium grows as underground stems and leafy, above-ground stems. However, the leafy shoots look nothing like any other liverwort and in some ways more like a moss. *Treubia* (with the related *Apotreubia*) is sort of in-between being leafy and thalloid, with a sort of ruffled-thalloid appearance. It was only when scientists looked at their DNA that the very different *Haplomitrium* and *Treubia* were found to be related. Having made that discovery, however, they then found other things that linked them together, such as the microscopic structure of their sperm.

The complex thalloid liverwort *Conocephalum conicum*. The pale dots are pores that open into air chambers inside the thallus. The prominent objects that look a little like mushrooms might be mistaken for sporophytes, but they are specialised structures produced by the thallus that merely support the sexual organs and the sporophytes, when they develop. In other complex thalloid species, such as the very common *Marchantia polymorpha*, these structures have projecting fingers and look a little like umbrellas. Image: © Des Callaghan

'Complex' thalloid liverworts

The thalloid liverworts most people are familiar with in their gardens and in other disturbed places are the so-called 'complex' thalloid liverworts. Close-up, these have a texture that looks a little like snakeskin, with little diamond-shaped 'scales', each one with a tiny pale dot in the middle. They are not scales of course – what you are seeing is the top surfaces of lots of little enclosed air chambers inside the plant. The pale dots are pores, entrances to these air chambers from the outside world.

It's because of this complicated structure of the thallus with its air chambers that we call these plants complex thalloid liverworts. The air chambers have lots of green filaments inside them that increase the surface area and allow gases for photosynthesis to be exchanged more easily, with carbon dioxide entering through the pore at the top. This is rather similar to how the leaves in more familiar plants work with their stomata and internal spaces, and less like the usual way that bryophytes work, absorbing gases and water over their entire external surfaces.

Many complex thalloid liverworts have specialised structures on which the sexual organs are produced that look like tiny, fingered umbrellas. The sporophytes are also produced on these, but the umbrella structures themselves are not the sporophytes, despite to some extent resembling 'blobs-on-sticks'!

What you are seeing is the top surfaces of lots of little enclosed air chambers inside the plant. The pale dots are pores, entrances to these air chambers from the outside world.

Plagiochasma rupestre, another complex thalloid liverwort with a stalked structure bearing the female reproductive parts. In this case these have been fertilised and the resulting dark, spherical sporophytes can be seen clearly. Image: © Des Callaghan

The simple thalloid liverwort *Moerckia flotoviana*. This species belongs to the general group of simple thalloid liverworts that often have fairly thick, rubbery-looking thalli, although in this species and its relatives the thallus is only a single layer of cells thick at the wide margins. The frilly structures in the middle are fringed scales that cover the male sexual structures. Image: Des Callaghan (CC BY-SA 4.0)

Simple thalloid liverworts

Strictly speaking (and very importantly for evolutionary biologists), this isn't actually a 'real' group! There are two groups of simple thalloid liverworts, but one of them is more closely related to leafy liverworts than it is to the other simple thalloid group. Nonetheless we will talk about them together here for simplicity.

As the name suggests, simple thalloid liverworts lack the complicated air chamber structure of complex thalloid liverworts, with the thallus being more or less uniform all of the way through. In the group that is most closely related to the leafy liverworts, the thallus is generally only a single layer of cells thick (except for a midrib at the centre), so the plants have the translucent appearance of leafy liverworts and mosses. Many of the species in the other group look a little more like complex thalloid liverworts with the thallus being several layers thick

(so rubbery-looking again), but without the snakeskin appearance created by the air chambers, although some have thalli that are only a single layer of cells thick in their outer parts.

The simple thalloid liverwort *Aneura mirabilis* (the 'ghostwort') lacks green pigments and gets nourishment from fungi that are connected to the roots of nearby trees. The smooth, cigar-like structures in this picture are young sporophytes, while the gametophyte thallus can also be seen (most prominently between the two leftmost sporophytes). Image: © Des Callaghan

The tropical leafy liverwort *Conoscyphus trapezioides*. Leafy liverworts have their leaves arranged in three rows, with the third row usually smaller and sometimes absent. This gives them a flattened look, or sometimes an almost caterpillar-like appearance if the leaves are folded forwards as in this species. Image: © Des Callaghan

Leafy liverworts

Just as with the mosses where the vast majority of species occur in the group with toothed sporophytes, the vast majority of liverworts (about 6,000 of the 7,000 species worldwide) are leafy liverworts, with stems and leaves quite like those of mosses. Earlier, when talking about how to tell mosses apart from leafy liverworts, we mentioned that leafy liverworts always have their leaves arranged in two or three straight rows, or *ranks*.

The leafy liverwort *Trichocolea tomentella* growing in Wales. This species has leaves that are extremely finely divided until they are almost hair-like at the ends, giving the plant a uniquely fuzzy appearance. Image: © Des Callaghan

Gottschelia schizopleura, a leafy liverwort
growing in tropical cloud forest on La Réunion.
Image: © Des Callaghan

Generally, there are two prominent ranks of leaves, one on each side of the stem, giving the liverwort a flattened or folded look. The third row of leaves, if it is present, is underneath. These 'underleaves' usually look different from the side leaves and are generally much smaller and less obvious.

Many of the differences between leafy liverwort species are in their side leaves and the ways these are arranged. They can have quite complicated shapes, often with one part of the leaf being doubled-over (folded backwards) to form what looks like a smaller leaf on the back side of the plant (although it is actually just part of the same leaf). These smaller *lobules* on the reverse side of the main part of the leaf are developed into extraordinary flask-like structures in some species. These capture and retain water, and may even trap tiny, microscopic animals that the liverwort could gain nutrition from. We will talk further about the possibility of carnivorous liverworts in the next chapter!

Generally, there are two prominent ranks of leaves, one on each side of the stem, giving the liverwort a flattened or folded look.

HORNWORTS

As we mentioned earlier there are only about 200 species of hornworts worldwide, so we won't go into detail about the different types. A very interesting feature of many hornworts is the presence of microscopic structures inside their cells called *pyrenoids*. These are special microscopic machines for increasing the concentration of carbon dioxide around the *chloroplasts*, the structures that all green plants have that allow them to make food through photosynthesis from water and air. Pyrenoids appear to have evolved several times within hornworts, with some species having them and others lacking them. It's possible that pyrenoids could have evolved during a period of the earth's history when carbon dioxide levels in the air were much lower than they are today.

Hornworts also have special associations with blue-green bacteria that live inside their thallus tissues. These bacteria are very good at capturing nitrogen, which can be in short supply in many of the places where hornworts grow and which the hornwort can then absorb. Interestingly, similar associations with blue-green bacteria have evolved in the complex thalloid liverwort *Blasia*.

There are only about 200 species of hornworts worldwide.

The leafy liverwort *Biantheridion undulifolium* growing in between the bright red stems of *Sphagnum rubellum*. Image: © Des Callaghan

Why mosses matter

Mosses and Us

It's unfortunate that many people are only aware of mosses as a substance that they would like to get rid of, often without really knowing why. Mosses grow uninvited on rooftops and between paving slabs, rudely disrupting the sterile uniformity of these artificial surfaces that are generally too inhospitable for other plants. We don't eat mosses or obtain raw materials such as wood or fibres from them, they don't lend themselves particularly well to horticulture (Japanese moss gardens notwithstanding), and compared to other plants they don't produce a particularly wide variety of novel chemical substances. So, what use are they?

In fact, quite apart from their intrinsic beauty and the homes they provide for a range of tiny animals, mosses play a critical role in climate change prevention and in beneficially controlling the way that water moves through forests, uplands and stream valleys. If mosses ceased to exist tomorrow, we would all be in a lot of trouble!

Many of the ways that mosses directly benefit us are related to their capacity for holding and controlling water.

> Mosses play a critical role in climate change prevention and in beneficially controlling the way that water moves through forests, uplands and stream valleys.

SPHAGNUM BOGS

Peat mosses (*Sphagnum*) are almost wholly responsible for creating and maintaining *peat*. Peat is used in gardening, as a traditional fuel used especially in Scotland and Ireland, and for the flavour it imparts to many whiskies. Peat burns because it is made of organic matter (mostly dead *Sphagnum* moss) that has not decomposed. Normally, dead organic matter decomposes very quickly, a process that results in the carbon of which it is largely made being released into the atmosphere as carbon dioxide. Peat bogs are highly unusual, because the dead *Sphagnum* moss doesn't decompose, but simply builds up gradually underneath the living surface layer. This happens because of the unique properties of the *Sphagnum* itself (see p. 58).

In simple terms, *Sphagnum* is able to keep the soil in which it grows permanently wet, largely preventing decomposition. The result of this is that about 20% of the total carbon stored on land in natural habitats is in the form of peat, i.e. dead *Sphagnum*! This makes it crucially important to conserve *Sphagnum* bogs, because it's the living surface layer of moss that largely prevents the peat from decomposing and releasing all of that carbon into the atmosphere, where it would act to accelerate climate change.

About 20% of the total carbon stored on land in natural habitats is in the form of peat, i.e. dead *Sphagnum*!

CAPTURING WATER AND CONTROLLING ITS MOVEMENT

Mosses are most abundant and most diverse in places where lots of water is available. This is because of the particular relationship that mosses have with water, absorbing it over their entire surface rather than from roots in the soil. The water can come from various sources, from rain as well as from the air in the form of mist or fog.

In places such as the misty forests found on mountain tops in the tropics (*tropical montane cloud forests* – see p. 118) and the wet forests found on the coastal fringes of cooler regions (*temperate rainforests* – see p. 152), great abundances of mosses and liverworts often clothe tree trunks in spongy layers many centimetres thick.

On tropical mountains these mosses capture large amounts of water from the clouds that constantly roll over the summits and then gradually release it into the streams that feed rivers lower down. At the same time, in these places and in cooler regions, mosses in forests and in moss-dominated bogs and moorlands gently control the flow of heavy rainfall, absorbing it like a giant sponge and then slowly letting it out again into rivers in a regulated manner. This has a very important role in flood prevention and the protection of soil from erosion. Think of it like a sink with a small plughole and a powerful tap – without the mosses the tap would be on full blast for short periods and off the rest of the time,

---

Mosses absorb water over their entire surface rather than from roots in the soil.

Mossy temperate rainforest in New Zealand.
In these types of forests mosses and their relatives
play an important role in controlling the movement
of water, due to their abundance and ability to
rapidly absorb moisture. Images: © Neil Bell

causing the sink to periodically overflow. With the mossy sponge absorbing sudden rainfall and then releasing it slowly, it's as if the tap is continuously on but at a much more gentle rate, allowing the water to flow down the plughole without filling up the sink.

Many mosses grow directly on bare rock or gravel, and so can play an important initial role in soil creation. This is *Pelekium versicolor*, growing on a boulder in a plantation on La Réunion.
Image: © Des Callaghan

Until recently this species, *Roaldia revoluta*, was called *Hypnum revolutum*. DNA studies have shown that the features that characterised *Hypnum* have evolved independently in species that are not very closely related, belonging even in different families. This has reduced the number of species of *Hypnum* from more than 40 to less than 10. In fact, the name *Hypnum* was originally applied to nearly all creeping, profusely branched mosses and the genus has been getting progressively smaller ever since. Image: © Des Callaghan

A CRITICAL PART OF NUTRIENT CYCLES AND FOOD CHAINS

Just as forests of trees are stalked by tigers and grazed by squirrels, so the miniature forests of mosses offer hunting grounds, protection and food for a host of much smaller (and often much weirder!) creatures. We will look at this more fully in the next section. Mosses and their relatives are very much involved in the cycling of nutrients that other living things are dependent on. They can have chemical effects on soils, either by directly capturing substances from the atmosphere or obtaining them with the help of fungi or bacteria, increasing the levels of important nutrients such as nitrogen and phosphorus. Also, because they are often the first plants that are able to grow on bare surfaces such as rocks or gravel due to not having roots, they play an important role in limiting erosion and in soil creation, paving the way (literally, in a sense) for other plants.

HISTORICAL USES OF MOSSES

Mosses have been used for stuffing pillows (which is why the name of some of them, *Hypnum*, refers to *hypnos* or 'sleep' in Greek), as a natural packing material and widely in horticulture and floral displays. Perhaps one of the most celebrated historical uses has been the employment of *Sphagnum* as a wound dressing, due to its exceptional absorbent qualities and its capacity to maintain acidic conditions that inhibit bacterial growth (see above, and discussion in Wetlands, p. 143). Documented in medieval Irish sources, it was subsequently used extensively in the First World War, with large quantities being gathered for the purpose in Scotland under the partial direction of Isaac Bayley Balfour, Professor of Botany at Edinburgh University and Regius Keeper of the Royal Botanic Garden Edinburgh.

Denizens of the Moss Forest

It's often the case in nature that very similar patterns of relationship are repeated at different scales – the same fundamental interactions that occur between different types of organisms on large scales also happen on much smaller scales. Frequently this happens in the same place, with *habitats* containing functionally similar *microhabitats* within them. Such is the case with mosses. Trees in forests populated by grazers and predators have mosses growing on them, and these mosses in turn form miniature forests populated by their own tiny grazers and predators.

Of course, size matters – the laws of physics and chemistry are the same everywhere, but they don't always operate in the same way at different scales. The most obvious example of this, which we touched upon earlier, is that small objects have a bigger surface area than large objects of exactly the same shape, *relative to their volume*. To see why this is the case, imagine cutting a cube into four smaller cubes. The volume of one of the smaller cubes will be a quarter of the volume of the original cube, but its surface will be *twice as much* as a quarter of the surface of the original cube, because it will include that quarter, plus the newly exposed surfaces created by cutting.

This is why most mosses are able to live by absorbing water and nutrients directly over their surfaces, and why they dry very quickly and

Trees in forests populated by grazers and predators have mosses growing on them, and these mosses in turn form miniature forests populated by their own tiny grazers and predators.

become wet again quickly – they have much bigger surfaces (relatively) to lose and gain water over. It's also why the animals we find in moss forests don't look like tiny tigers and tiny lizards! You don't need lungs to breathe if you are very small, for example, and being constantly warm blooded would be almost impossible for something the size of a mite.

WHAT ANIMALS ARE COMMON IN MOSS?

If you look at a clump of fresh moss under a low-power microscope, you will often find that it is teeming with tiny 'critters'. Most of these animals are probably using the moss as a place to find shelter, graze fungi or hunt other animals, reproduce and lay eggs. There are some that consume the moss itself for food. At the scale of what can be seen with the naked eye or just a little bit smaller, you will see many small worms, mites and various small insect relatives such as springtails and coneheads. Larger animals such as small spiders and beetles are also often seen.

For animals the size of mites and springtails, the moss plants are genuinely a jungle – a vast space of giant stems and leaves to be climbed in search of food, and with dangerous predators to be avoided. But there is also an abundance of (usually) slightly smaller animals that we can only really see under a microscope – in particular, the marvellous and surprising rotifers and tardigrades. These creatures are living within the water films and droplets that are a semi-permanent feature of most mosses.

These creatures are living within the water films and droplets that are a semi-permanent feature of most mosses.

To many microscopic creatures, a clump of
moss is a vast forest in which to live and hunt!
(This is *Pterigynandrum filiforme* in Iceland.)
Image: © Des Callaghan

An oribatid mite (*Phauloppia lucorum*) on moss.
Image: © Andrew Murray

MITES AND SPRINGTAILS

These are perhaps the most prominent creatures at the millimetre to sub-millimetre scale in many clumps of moss. Mites are related to spiders but are unsegmented and smaller (usually less than one millimetre long). The particular mites that are most common in mosses are *oribatid mites*, sometimes called 'armoured mites' because they are entirely encased in hard shells like suits of armour. This rather effectively protects them from predators, such as beetles and spiders. Most of them graze on fungi that are found growing on the moss, although some hunt tiny worms. There are hundreds or even thousands of different species found on mosses worldwide, each one adapted to a very particular way of living. They may have quite an easy life – their food is readily available

These are sometimes called 'armoured mites' because they are entirely encased in hard shells like suits of armour.

The springtail *Dicyrtoma fusca*. These creatures are not insects, although they are six-legged and closely related to insects. Image: © Andrew Murray

in the moss forest, and one study has suggested that there are not many other animals able to eat them!

Life is perhaps a little harder for the springtails, which are known to be eaten by many other animals, including small poisonous spiders, beetles and large mites – as well as other springtails. In turn, they may eat smaller animals such as worms and rotifers, but like mites, many are grazers and mostly eat small fungi. Like the rather unassuming coneheads, they were once regarded as primitive wingless insects but are now thought to be a separate group, although related to insects. Springtails get their name from their possession of a unique structure, resembling a large double prong, that is folded under their bodies and effectively spring-loaded. When released this enables them to jump up to 20 times their own body length. The principle is a little like an upside-down mouse trap!

A tardigrade, or 'water bear'.
Image: iStock.com/fruttipics

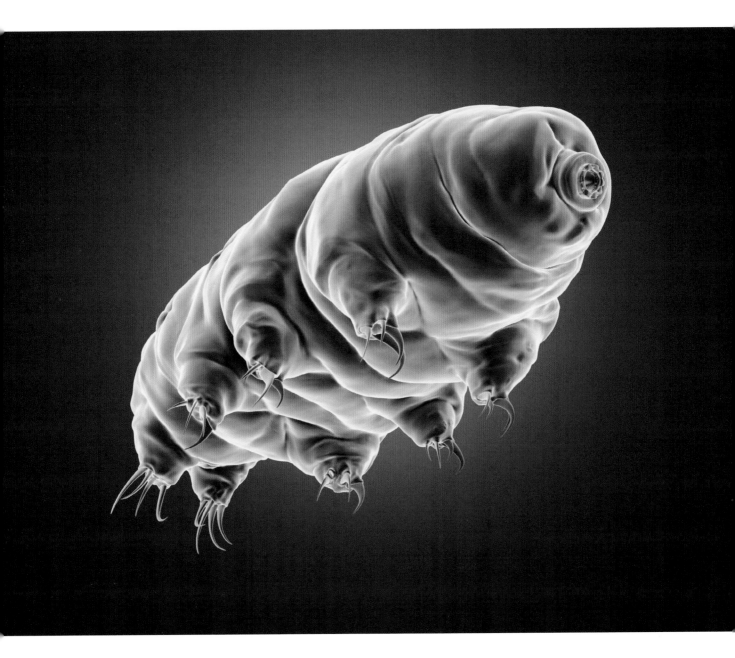

TARDIGRADES

Tardigrades are the celebrities of the moss-animal world, with a combination of cuteness and an amazing resilience to extreme conditions that has attracted occasional media attention in the last few years. Fans of *Star Trek* will be familiar with the giant, sentient, space-dwelling tardigrades in *Star Trek: Discovery*. While real tardigrades don't navigate the multiverse and provide formidable adversaries for klingons, they have been shown to be able to survive the vacuum and intense UV radiation of space in a dormant form!

Tardigrades are an exception to the rule we mentioned earlier whereby animals at different scales generally look different. They are often called 'water bears' or 'moss piglets', because they look rather like chubby little bears complete with clawed paws (although with eight legs rather than four). They are members of their own group that is related to, but quite separate from, the big group (*arthropods*) that includes all of the insects, spiders, crustaceans and centipedes (as well as mites and springtails). They are really rather unusual!

Their ability to survive in space is down to being able to enter a dormant state, called a *tun*. In the tun state, the tardigrade is almost completely desiccated, so body processes are halted almost entirely (life can't 'work' without water, because its working parts can't move and interact without it). In this state they are able to survive freezing, boiling, vacuum, extreme pressures and any number of poisons, but are able to rejuvenate and start moving and feeding again once rehydrated.

Of course, in this sense, tardigrades play exactly the same role in the animal world as mosses do within the world of plants. Because most of them live on mosses and lichens and don't resist drying out in the way that other small animals like mites do, they too have to 'go with the flow' of constant desiccation and rehydration that mosses are adapted to live by. And they probably do it in a similar way – using special proteins as well as chemicals with 'anti-freeze' properties to protect their bodies at a microscopic level as they slowly dry out. It's perhaps a little like erecting a protective molecular scaffolding to replace the water molecules (which normally have that role) as they gradually disappear. And of course, once in this 'packaged up' condition, they are more able to resist other forces that would be damaging to life in its active state.

ROTIFERS

Rotifers, like tardigrades, are very small multicellular animals that exist in films of water and are more or less transparent. Despite being multicellular and having a digestive tract like 'proper' animals, they are really existing in the world of single-cell organisms such as bacteria and protists (protists are single-cell organisms that have a nucleus, unlike bacteria). And indeed, it is these single-celled lifeforms that are the rotifer's main prey.

'Rotifer' means 'wheel bearer' in Latin, referring to the hypnotising, apparently whirling wheels that rotifers appear to have at one end of their bodies. In fact, these 'wheels' are really rings of hairs that beat in a very

fast and rhythmic way, drawing water into the mouth of the rotifer. This water contains dead organic material as well as bacteria and protists that the rotifer feeds on. Sometimes rotifers will be seen swimming through the film of water that clothes mosses in search of food, while other times they will anchor themselves to a single spot and feed from there.

There are a number of different groups of rotifers each with their own characteristics, and a great number of species, many of which have still to be discovered because many species look the same as other ones.

Just like tardigrades, rotifers can survive as almost completely desiccated *tuns*, and they have also been shown to be able to survive in space!

> Just like tardigrades, rotifers can survive as almost completely desiccated *tuns*, and they have also been shown to be able to survive in space!

A species of rotifer from the genus *Nothocla*.
To feed, these tiny (but nonetheless multicellular)
animals draw water into their mouths by beating
a 'crown' of hairs very quickly. These hairs can be
seen at the left side of the animal in this picture.
When moving, they create the illusion of miniature
wheels spinning. Image: Wiedehopf20 (CC BY-SA
4.0), with cropping

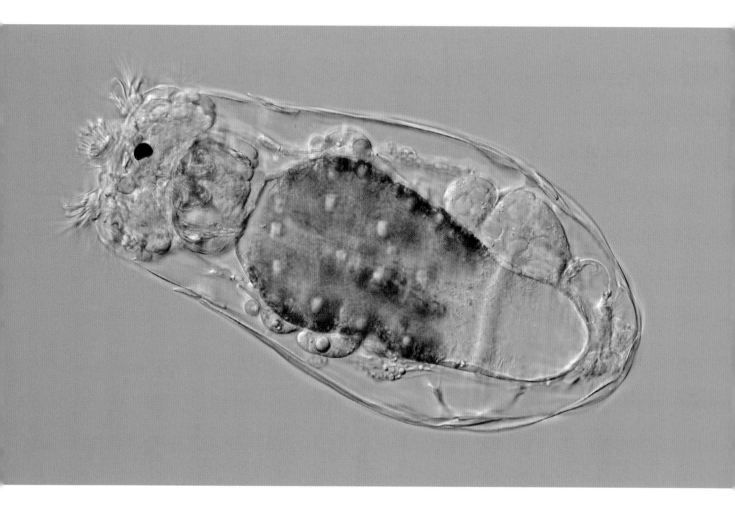

Mosses with long creeping stems and abundant branching (sometimes informally called feather mosses) make great nesting material, as this wren knows. Image: © John Queen

Other Interesting Animal–Moss Interactions

Mosses and their relatives are a prominent and abundant part of many habitats and interact with many of the animals there in a variety of ways. As well as creating living spaces for tiny animals they provide raw materials for larger ones, and sometimes even use animals for dispersal of their spores. There is even evidence that some liverwort species are effectively carnivorous – trapping single-celled protists and other tiny creatures and absorbing nutrients from them, just like other more familiar carnivorous plants such as sundews and pitcher plants.

Different, closely related bird species living in the same forest select different species of moss to make their nests, based on the size of their broods.

NESTING MATERIAL

Mosses are very commonly used as material for building nests by birds. The long, feathery stems of some types of forest moss are ideal for weaving between twigs and other materials to create a solid structure. Furthermore, the moss may provide insulation (think of it like the woolly insulation in your loft!), as well as perhaps controlling humidity by absorbing moisture from inside the nest as well as from the surroundings. Remarkably, Tomasz Wesołowski and Sylwia Wierzcholska recently demonstrated that different, closely related bird species living in the same forest in Poland select different species of moss to make their nests, based on the size of their broods – species with more chicks chose more robust moss species to make their nests, and knew how to find them.

Many mosses in the family Splachnaceae grow on dung or carrion and attract flies to disperse their spores. The lower part of the spore capsule is often expanded to form a landing platform for insects. The one in this species, *Splachnum rubrum*, is perhaps the most spectacular! Image: © Michael Lueth

SPORE DISPERSAL BY INSECTS

One of the most dramatic interactions between mosses and animals occurs in a particular family of mosses called the Splachnaceae. In this family many species actively attract insects to disperse their spores. In nearly all mosses, spore dispersal is by wind (or occasionally water). This contrasts with flowering plants, where many species use fruit to attract animals to disperse their seeds, while of course most flowering plants attract insects to disperse pollen.

This moss family contains many species that grow on animal dung or carrion. These substances occur temporarily and rather unpredictably, so if you are a dung moss, making sure that your spores find the next available bit of dung to

Although the fly-attracting sporophytes of *Tayloria gunnii* (left) and *Splachnum vasculosum* (right) look rather similar, they have evolved independently in Tasmania and the northern hemisphere respectively. Images: © Lepp, H. (left) and © Des Callaghan (right)

grow on potentially poses a problem. The solution is to find an animal that is mobile and good at finding dung, and to persuade it to carry your spores! It seems that growing on dung or carrion has evolved together with attracting insects for spore dispersal, and this combination of characteristics has in turn evolved several times within the family.

These mosses attract dung flies by creating a landing platform for them on their sporophytes, which is often brightly coloured. They also often exude chemicals that particular dung flies can smell and are attracted to. Their spores are large and sticky, so when a fly lands on a sporophyte it is likely that some

spores will stick and be transported to the next cowpat or reindeer dropping. A lovely demonstration of how evolution often finds the same solution twice can be seen in the rather similar types of 'landing platform' that have evolved independently in relatively distantly related species within the family, such as *Tayloria gunnii* in Tasmania and *Splachnum vasculosum* in the northern hemisphere arctic and subarctic.

These mosses attract dung flies by creating a landing platform for them on their sporophytes, which is often brightly coloured.

CARNIVOROUS LIVERWORTS!

It's well known that some flowering plants growing in wet, nitrogen-poor environments have evolved to trap and digest animals (usually insects) in order to gain nitrogen and other nutrients. Most people know of the Venus flytrap, while in bogs (including in the UK) the sundew plant uses sticky hairs to trap insects and then special enzymes to digest them. Is it possible that some liverworts could be doing exactly the same thing except at a smaller scale, trapping and 'eating' single-celled protists instead of insects?

We mentioned earlier that some liverworts have the rear parts of their leaves, or *lobules*, shaped into tiny flask-like structures. In the liverwort *Frullania* these look a little like tiny helmets, and they have been commonly thought to function as a method of slowing drying of the plant by trapping and retaining water. However, there are a couple of types of liverwort that have taken this process a little further.

Colura and *Pleurozia* (both of which are present in the UK) are rather distantly related, but they both

Some liverworts could be doing exactly the same thing except at a smaller scale.

have parts of their leaves developed into enclosed flasks that have little flaps at their entrances. Because these flaps are closed when the plant is wet and open when it is dry, one theory is that they act to trap water when conditions are dry and retain it when it is wet. But another possible function is hinted at by the way that the flaps effectively act as one-way valves when the plant is wet. It has now been experimentally demonstrated in both *Pleurozia* and *Colura* that protists can push on these flaps to gain entry into the lobule or water sac, where they become trapped and eventually die. All that would then be needed for the liverwort to gain nutrients from the protists would be for them to be digested by enzymes released by the plant or by bacteria that are also trapped in the water sac. While it has not been scientifically demonstrated that this happens, it seems plausible that it could do and that these liverworts are true carnivores!

Protists can push on these flaps to gain entry into the lobule or water sac, where they become trapped and eventually die.

Species of *Pleurozia*, such as this tropical *Pleurozia gigantea*, may trap and digest microscopic organisms in their flask-like leaf lobes. One species, *Pleurozia purpurea*, is quite common on wet heath in the west of Scotland. Image: © Des Callaghan

Ptychomitrium polyphyllum, a common moss forming cushions on rocks in the wetter parts of Britain, especially in coastal areas.
Image: © Des Callaghan

Moss habitats

The 'sword moss' *Bryoxiphium norvegicum*
growing on the ceiling of a cave in Iceland.
The English name refers to the sharply flattened
shoots with two ranks of closely-spaced leaves.
Image: © Des Callaghan

From Rainforests to Tundra

Mosses and their relatives are found nearly everywhere in the world, from arctic tundra and the tops of snow-covered mountains to the hottest deserts and the cracks between paving slabs in polluted city centres. In places such as wet forests and bogs they can be abundant, covering every available surface with luxuriant wefts, cushions and carpets. In other places they are present but barely visible, existing as tiny, frequently dried-out tufts in isolated crevices. Mosses, of one species or another, are *able* to grow in many places (and often in places where other plants can't), but they are usually most luxuriant and have the highest numbers of species in places that have a high availability of moisture, particularly atmospheric moisture, throughout the year.

Perhaps surprisingly, mosses and their relatives are not particularly prominent or diverse in *lowland* tropical rainforest – the sort of dense, very tall forest that most

people have in mind when they think of 'rainforest'. This is because these forests have a very complex structure dominated by many layers of flowering plants that are able to thrive in the stability that these places provide. These plants use up almost every bit of light so that the forest floor itself is a very dark place covered in leaf litter that most mosses find rather inhospitable. Nonetheless, some mosses can be rather abundant in parts of the upper canopy, where light levels are higher but not too extreme, and they have plenty of different surfaces to grow on.

Mosses and liverworts are really in their element in cooler rainforests with shorter, more open canopies. These types of rainforest are found at higher altitudes in the tropics and in coastal areas near sea level in temperate regions. Both of these habitat types, *tropical montane cloud forest* and *temperate rainforest* respectively, are rare throughout the world as a whole, and conserving them is very important for conserving the diversity of mosses.

Mountains and wetlands (of various types) are other places where mosses and liverworts are often diverse and luxuriant. Indeed, in many of these places the usual botanical state of affairs is reversed. Mosses and their relatives are the dominant plants, with other plants reduced to a subsidiary role and barely clinging on. This is the case on exposed, rocky mountain tops with little in the way of the soil that most other plants need to grow, or in places that are covered with snow for most of the year, such as sheltered spots high in the mountains or in the Arctic and Antarctic regions. And then there are the very wet (if rarely truly aquatic) places that are either too damp and too acidic for other plants to thrive in, such as *Sphagnum* bogs, or too rocky and disturbed, such as the rocks in and around fast-flowing streams.

Because many species of moss often do well in disturbed areas, humans can often unintentionally give them a helping hand. Gravelly paths, lime-rich walls and concrete, mine spoil heaps and stubble fields can provide places for particular species of moss to grow. Some of these species may be very rare in natural places because they have very exacting requirements, such as copper or lead in the soil, or a particular type of regular disturbance that humans can provide.

Mosses and their relatives are *different* from other plants, taking advantage of these differences to exploit places that they are much better suited to (just as other plants do in the places to which they are more suited). On the other hand, they are still plants, so they are often competing for the same places to grow, in an uneasy equilibrium that can be shifted one way or the other by small changes that favour or disadvantage them compared to other plants.

Mosses and their relatives are able to grow in many places where other plants can't. *Grimmia montana* and *Grimmia decipiens* growing together with lichens on volcanic rock in Scotland.
Image: © Des Callaghan

Tropical Cloud Forest

On the upper slopes of mountains across the tropics, a type of cool, wet forest is found that provides one of the best habitats on earth for mosses and their relatives to grow in – so much so that it is sometimes referred to as 'mossy forest', even by botanists who don't normally pay much attention to mosses! Its other name is 'cloud forest' – referring to the almost constant presence of low cloud from weather systems that move over the mountains, causing atmospheric moisture to condense at high altitudes.

The trees in these forests are often almost completely covered in mosses and liverworts, which clothe the twigs, trunks and branches in a thick, fuzzy blanket. Mosses often dominate on the forest floor too, where dead wood builds up in the hollows between rocks and boulders and everything is covered by a spongy mat of moss, creating something more like a three-dimensional mossy maze than a 'floor' or a carpet. These are magical places teeming with life where mosses play a fundamental role, soaking up moisture and nutrients from the air, storing them within the forest and then channelling them down

---

These are magical places teeming with life where mosses play a fundamental role, soaking up moisture and nutrients from the air.

The moss *Orthostichella versicolor* growing on a tree trunk in tropical cloud forest on La Réunion. Quite a few wet tropical forest mosses have strongly concave leaves that are able to retain water when the plant is wet, almost like hundreds of little cups. Image: © Des Callaghan

Phyllogonium fulgens, a moss that typically forms long, dangling shoots hanging from trees in tropical forests. This particular example is on an earth bank beside a path in tropical cloud forest on La Réunion. Image: © Des Callaghan

◀ The large moss *Euptychium* growing profusely on a young tree sapling in cloud forest in New Caledonia. Image: © Neil Bell

▼ Tropical cloud forest on Mt. Aopinie, New Caledonia. The nearly constant cloud condensing on the upper slopes ensures that the interior of the forest is dominated by an abundance of mosses. Image: © Neil Bell

the slopes to feed the streams and rivers in the valleys. They function, in effect, like a forest-wide system for harvesting water and nutrients from the clouds.

These mossy cloud forests are found in three main places in the world – the South American Andes and Central America, the mountains of the large South East Asian islands such as Borneo and Papua New Guinea, and (to a lesser extent) the sub-Saharan African mountains. They are also found in Eastern Queensland in Australia (where they share a lot in common with the forests of New Guinea to the north to which they were once connected) and on some of the smaller Pacific Islands. The species and even the families that are found in each of the three main regions are generally different, and each type of cloud forest has its own unique character.

◄▶ *Dawsonia* – the largest upright-growing moss on the planet! Images: By Natalie Johnston (CCO) (left) and By Logan Lucas (CCO) (right)

GIANT MOSSES

Because the forest is sheltered, continually wet and stable over many years and through the seasons, some of these mosses are able to grow very large – large for mosses of course. Some species growing on tree trunks can produce stems more than 30 centimetres long, while a few that grow on the ground can be up to 60 centimetres high.

The biggest mosses are found in the South East Asian cloud forests, such as those in Papua New Guinea. The largest upright-growing, self-supporting mosses are species of *Dawsonia*. These are in the same family as the mosses that have the 'pepper-pot' style of peristome (pp. 51, 64), except that in these large species the peristome resembles a miniature brush, with the spores released from between the bristles. The purpose is the same – ensuring that the spores are released gradually over a long time (and generally when conditions are dry) rather than all at once.

Dawsonia, like most of the other species in the Polytrichaceae family to which it belongs, is very unusual among mosses and their relatives in having a well-developed *vascular system* of internal tubes to move water and nutrients around the plant. These mosses also have leaves that are more than a single layer of cells thick, so they don't have the translucent appearance of most mosses. In fact, many mosses in this family look strikingly similar to young pine seedlings! The complicated shape of these thicker leaves provides a greater surface area for exchange of gases for photosynthesis, in a similar-but-different way to the thick leaves of more familiar plants.

Remarkably, it was recently shown (by Timothy Brodribb and

The giant moss *Spiridens spiridentoides* growing on a tree in New Caledonia. This particular species is only found on one mountain top. Image: © Neil Bell

collaborators) that mosses in this family are probably able to *transpire*, just like other plants do. This means that evaporation of water from the leaves 'pulls up' water from lower down the plant stem through the vascular system by capillary action. *Dawsonia* doesn't have roots, but otherwise it functions a lot like more familiar plants that do, and much less like most other mosses. This of course largely explains why it is able to grow so large – it's much less limited by the need to be able to rapidly absorb water over its entire surface (p. 21), which becomes progressively more difficult the larger a plant gets.

They form a fuzzy covering over a foot thick that can almost dwarf the trunks of the trees they grow on.

Another group of giant mosses found in South East Asian cloud forests grow as long, horizontal stems attached to tree trunks. This is *Spiridens* – a group that is only found in tropical parts of South East Asia, Australasia and Oceania. Species of *Spiridens* are often so robust and abundant that they form a fuzzy covering over a foot thick that can almost dwarf the trunks of the trees they grow on. Meanwhile other mosses, in the family Meteoriaceae, have much more slender stems but are extremely long and dangly, forming pendent curtains hanging from horizontal branches and twigs.

TINY LIVERWORTS AND LIVERWORT-LIKE MOSSES

At the other extreme, hundreds of different species of tiny liverworts (and some mosses) can be found growing on twigs and on the thick, evergreen leaves of the trees – and sometimes even on the stems and leaves of larger mosses. Liverworts in particular may be well suited to this lifestyle because of their 'flattened' structure due to having their leaves in two main ranks, rather than spirally arranged as in most mosses. This allows very small species to grow closely pressed to the flat, leathery surfaces of evergreen leaves and helps to retain moisture. The diversity of these plants in many areas is rather poorly known, and it's likely that many species have still to be discovered.

Most of these liverworts growing on other plants (so-called *epiphytes*) are leafy, and the large majority belong to one major group of leafy liverworts (the taxonomic order Porellales). Fascinatingly, studies of the evolutionary history of leafy liverworts suggest that this group suddenly evolved more species just at the time when flowering plant forests were evolving, from around 100 million years ago onwards. Before this time, forests were mostly composed of conifers, possibly with many fewer suitable surfaces for liverworts and mosses to grow on. The evolution of flowering plant forests may have stimulated the evolution of epiphyte liverworts in the order Porellales. By contrast, most species in the other large group of leafy liverworts, the order Jungermanniales, are not epiphytes and this group didn't suddenly become more diverse at the time that the Porellales did.

Among the mosses, there is one large group that is particularly

Epiphyllous liverworts growing on the large leaf
of an evergreen tropical tree in Central America.
Image: © Scott Zona

The liverwort *Lepidolejeunea delessertii* growing on rotting bark in lowland tropical forest on La Réunion. Image: © Des Callaghan

Pleurocarpous mosses

When describing the major groups of mosses (see p. 50), it was mentioned that the vast majority of moss species occur in the group that has 'toothed' sporophytes of the 'active' type – in other words, spore capsules in which the gradual release of the spores is controlled by tooth-like structures that open when conditions are best for dispersal (usually when it is dry). Of the species in this group, around half are *pleurocarpous.*

Pleurocarpous mosses usually have more complicated branching than other mosses. Rather than having simple, short, upright-growing stems, they tend to have longer stems that grow more horizontally and are often regularly branched. They may look like ferns or tiny feathers, with many short side branches growing sideways from a central stem, or even like miniature trees with tufts of branches atop tiny 'trunks'.

This diversity of more complicated branching structures in pleurocarpous mosses appears to have come about because this group has found a way, through evolution, to stop reproduction interfering with the growth of long, branched stems. When a moss reproduces to produce sporophytes (see p. 25) this always happens at the end of a stem or branch, and it prevents that stem or branch from growing any further. In pleurocarpous mosses the reproductive structures are produced on extremely short side branches that exist for this purpose alone, rather than at the ends of the main stems or branches. This means that the main stems can keep on growing and branching, even when they are also producing reproductive structures and sporophytes. In pleurocarpous mosses it looks as if the sporophytes are attached to the sides of the main stem rather than the stem tips, but they are actually attached to the tips of these tiny, specialised side branches.

The pleurocarpous moss *Exsertotheca intermedia*. Pleurocarpous mosses are fertilised and produce sporophytes at the ends of specialised short branches, rather than at the ends of main shoots or normal branches. Because these specialised branches are very short it often appears as if the sporophyte is growing out of the side of the main shoot itself. However, in this particular species we can see the specialised fertile branch very clearly because its leaves grow very long after fertilisation, sheathing the base of the sporophyte. Image: © Des Callaghan

associated with cloud forests in the tropics. With around 1,000 species worldwide, the taxonomic order Hookeriales is the second most diverse group of *pleurocarpous* mosses (see Box opposite), although the vast majority of species are only found in tropical forests. In Britain there are only three species that belong to this order, and in most European countries there is only one, or none at all. The fact that we have even as many as three species from this tropical rainforest moss group in Britain reflects similarities between the mild, continually wet climate of our coastal areas and the high altitude mossy rainforests of the tropics.

Most species in the Hookeriales have flattened shoots, with leaves that look as if they are arranged either side of the stem rather than spirally as in most other mosses. In fact they *are* mostly attached spirally to the stem, but the shoot develops a flattened shape, superficially resembling a leafy liverwort. Probably this flattened form helps the moss retain water when growing closely pressed to another plant that it is on. This seems to be the case for leafy liverworts that grow on other plants.

They may look like ferns or tiny feathers, or even like miniature trees with tufts of branches atop tiny 'trunks'.

Three species from this moss group are in Britain, reflecting the similarities between the mild, continually wet climate of our coastal areas and the high altitude mossy rainforests of the tropics.

▼ The liverwort *Plagiochila terebrans* in tropical
cloud forest on the island of La Réunion. Some
species of *Plagiochila* in tropical and southern
temperate areas adopt a fan-shaped growth form,
which is unknown in British species of this genus.
Image: © Des Callaghan

▶ *Porotrichum madagassum* (top) and *Deslooveria usagara*, two fan mosses growing on trees in tropical cloud forest on the island of La Réunion. Images: © Des Callaghan

FAN MOSSES AND TREE MOSSES

As well as small, 'flattened' liverworts and mosses, there are other growth forms that are particularly common in cloud forests. Perhaps the most noticeable is the 'fan' shape, which is particularly associated with medium-to-large-sized mosses growing horizontally from tree trunks. Fan mosses are attached to the surface they are growing on by a single large stem, which then produces many short, closely spaced branches that are all orientated in a single plane. This allows the moss to have as much green, photosynthetic tissue as possible

Looking down on the 'canopy' of the tree moss *Hypopterygium tamarisci*. Image: © Des Callaghan

orientated towards the direction from which most of the light is coming, which in a forest is usually from above.

Strikingly similar forms have arisen independently in several different families of pleurocarpous mosses, and even in some liverworts (which are not usually large plants with regular branching). People familiar with European liverworts are often amazed to discover that species closely related to the small, simply branched ones they know, form large fans in tropical cloud forests!

Just as the fan shape is perfect for a large moss growing horizontally

> Mosses growing vertically on the ground or from other horizontal surfaces can benefit from adopting growth forms that resemble tiny trees.

from a tree trunk, similar mosses growing vertically on the ground or from other horizontal surfaces can benefit from adopting growth forms that resemble tiny trees. In this case there is also a plane of photosynthetic tissue orientated towards the light, but it is perpendicular to the direction of growth of the stem and raised clear of surrounding obstructions. Of course, full-sized trees adopt the same shape for exactly the same reasons. In many cases, species of moss that have this tree-like shape are closely related to other ones that have the fan shape, and it seems that not much is required for one of these growth forms to evolve into the other. Basically, the structure of a fan is quite like a flattened tree.

Fan mosses and tree mosses are found in temperate rainforests too (especially in the southern hemisphere), but they are particularly characteristic of tropical cloud forests.

While only a few mosses and liverworts are truly aquatic, many thrive in frequently inundated or permanently waterlogged places. Here, the complex thalloid liverwort *Marchantia polymorpha* subspecies *montivagans* is growing intermixed with the moss *Calliergon cordifolium* in standing water in wet woodland. Image: © Des Callaghan

Wetlands

Mosses and their relatives have a life strategy of tolerating rather than resisting drying out. As they are able to obtain water and chemicals over their entire surface and from the atmosphere rather than from roots, they grow best in places where water is usually available. Other types of plant often struggle to cope with permanently waterlogged soil, lack of soil, or physical disturbance from flowing water.

Ecologically, the most important mossy wetland habitats are bogs. As we've seen, the ability of *Sphagnum* moss to prevent decomposition in bogs allows peat to build up over hundreds of years, keeping carbon in the ground that would otherwise be released into the atmosphere. It's the unique properties of *Sphagnum* that allow it to do this (see pp. 58, 83); this is further described below.

Tolerant of salty sea-spray, the moss *Schistidium maritimum* is found on coastal rocks and is about the nearest thing to a marine moss that exists.
Image: © Neil Bell

Other places where mosses and liverworts can dominate include those that are very wet most of the time but usually not completely submerged, such as the edges of lakes and rivers, or places that are more or less permanently wet due to spray or run-off, such as rocks beside waterfalls and springs and flushes in the mountains. A particular situation in which mosses thrive is in or beside small, rocky streams in hilly country, where fast-flowing water keeps the surfaces wet and soil-free, but without these surfaces being permanently under the water. This is the clichéd mossy brook, babbling with the sound of water splashing over rocks on which mosses and liverworts have gained a secure foothold with their small, hair-like structures called *rhizoids*. Finally, there are some mosses and liverworts that do spend practically all of their lives fully submerged in fresh water. Some of these are always found in such situations, while others are normally land-dwelling species that are capable of living under water, although often they grow in completely different ways when they do so!

Mosses and their relatives appear to have evolved from freshwater algae, which may partially explain why there aren't any sea-dwelling (*marine*) mosses. Nonetheless there are some mosses that grow in saltmarshes, while others are able to grow very near the high-tide mark on rocks that are often deluged with sea-spray.

Sphagnum moss keeps the soil in which it grows permanently wet, preventing the dead organic material from decomposing.

SPHAGNUM BOGS

As we saw earlier, *Sphagnum* moss keeps the soil in which it grows permanently wet, thus preventing the dead organic material (which is mostly dead *Sphagnum* moss itself) from decomposing. This undecomposed material (peat) accounts for 20% of the total carbon stored in land in natural habitats.

All life on earth is based on carbon – it's physically made of complex carbon molecules (mostly proteins, carbohydrates and fats) that are in turn all made of chains of carbon atoms joined together, along with other atoms such as hydrogen, oxygen and nitrogen. It takes energy to join up these carbon atoms, and this energy is stored in the molecules, ready to be released when the carbon atom chains are broken up again. Animals and fungi get the energy they need to live from breaking up large carbon molecules, which at the same time as providing energy, releases single carbon atom molecules into the air as the greenhouse gas carbon dioxide. This means that the bodies of dead plants and animals are nearly always eaten or decomposed by fungi for their energy – because free energy lying around is always going to be taken and used by something!

The special structure of the leaves of *Sphagnum* with their two types of cells has already been described (p. 58), the non-green cells acting like hundreds of little water bottles to store 20 times the weight of the plant itself in water. This is what allows the surface of a *Sphagnum* bog

to remain constantly waterlogged. Another trick that *Sphagnum* uses is to actively pump *protons* (hydrogen atoms stripped of their electrons) into the soil. This takes energy, but it makes the soil much more acidic. The combination of being waterlogged and acidic makes the soil very inhospitable to other plants, which suits the *Sphagnum*. But it also makes it inhospitable for the fungi and other organisms that normally decompose dead plant material for energy, so dead *Sphagnum* simply

It also makes it inhospitable for the fungi and other organisms that normally decompose dead plant material.

▼ Bogs aren't just *Sphagnum*! The liverwort *Odontoschisma sphagni* is just one of the moss relatives that makes its home on *Sphagnum* hummocks. Image: © Neil Bell

▶ *Sphagnum* moss. This species, *Sphagnum skyense*, is only found in the British Isles. Image: © Neil Bell

builds up underneath the living surface layer of the bog to make peat.

As well as being a huge *carbon sink* (an accumulator of carbon from the atmosphere), *Sphagnum* bogs can also be a source of carbon. If the bogs become damaged or dry out, carbon is released into the atmosphere.

Climate warming could make this more likely to happen, resulting in the stored carbon being released through decomposition. This in turn would cause more warming and more drying. *Sphagnum* bogs are therefore a potential *positive feedback loop* for climate change – they could accelerate rises in global temperature. This is why it is so important to fully understand how they function and to protect them – and part of that understanding is to understand their ecology and diversity.

We've already seen that the living surface layer of the *Sphagnum* bog can be likened to a giant wet blanket. The individual *Sphagnum* plants are shaped to allow the whole community of *Sphagnum* stems to act together as one. The tops of the stems have lots of short branches compressed into a broad, dense head, and when the stems grow next to each other these heads are all squashed

together to form a continuous, sponge-like surface.

Different species of *Sphagnum* form a mosaic across the bog, with some species forming large humps or *hummocks* in the drier areas, while others make soggy carpets in the hollows in-between. *Sphagnum* is very colourful, with reds, pinks, oranges, browns and yellows (as well as different shades of green) found in different species, with sometimes more than one colour in the same species. The very wettest areas of the bog form pools, but even there

Sphagnum is found in the form of *Sphagnum cuspidatum*, the 'drowned kitten moss', so-called because of its appearance when pulled out of a pool!

Many plants and animals make their homes in *Sphagnum* bogs, including insects as well as other species of mosses and liverworts. The dense stems of *Sphagnum* themselves form a space for several species of small liverworts to grow in, such as the attractive *Odontoschisma sphagni*. And near the edges of some bogs where birch trees encroach, if you pull up a mat of *Sphagnum* you may just find the ghostwort, *Aneura mirabilis* (see pp. 72–73). This remarkable white, thalloid liverwort entirely lacks green pigments and gets its food from fungi that connect it to the roots of nearby trees, rather than from photosynthesising itself.

When the stems grow next to each other these heads are all squashed together to form a continuous, sponge-like surface.

Sphagnum and the moon

The extent to which the carpet of *Sphagnum* on a bog acts almost as if it was a single individual (a 'supraorganismic system') has been highlighted recently by some fascinating research by Victor Mironov showing that the growth of *Sphagnum* is strongly influenced by natural cycles in the environment, including the lunar cycle.

Not surprisingly perhaps, *Sphagnum* grows faster in the summer when the temperature is warmer, and it appears to be temperature itself (rather than day length) that controls this. But this annual cycle is one of at least three cycles that coordinate the growth of the individual stems in a *Sphagnum* bog.

The second cycle corresponds to the lunar cycle – specifically, the 29.5 days between one full moon and the next. However, the *Sphagnum* plants in fact grow more vigorously near the time of the new moon (when the moon is almost invisible), and less strongly when the moon is full. It's thought that the plants can detect the levels of light at night and use this to control growth, with moonlight inhibiting growth. By using external, universal cues such as moonlight, the growth of individual stems can be coordinated. This is important for the function of the bog, because a flat, uniform surface in which adjacent *Sphagnum* heads are all at the same height will be better at retaining water.

The third cycle is shorter (7–16 days) and a little mysterious, although it seems to be coordinated with the 29.5 day lunar cycle.

▼ The unusual trellis-like structure of the *Fontinalis* peristome is likely to be an adaptation to releasing spores in very wet environments. Image: Courtesy of University of British Columbia blogs.ubc.ca

▶ The aquatic moss *Fontinalis antipyretica* has strongly keeled leaves and three-sided shoots. Image: Hermann Schachner (CCO 1.0)

AQUATIC MOSSES

Arguably no mosses are completely aquatic, because even species that often spend nearly all of their lives submerged can also be found surviving out of water. However, there are a number of species that are usually found in water and are clearly adapted to this lifestyle. In Britain, our most clearly aquatic mosses are our two species of *Fontinalis*. Stems of *Fontinalis antipyretica* can be more than 15 centimetres long and have strangely keeled leaves, giving the shoots a sharply three-sided appearance. Sporophytes are rare, but when they do occur the *peristome* (see image below) has its narrow inner teeth formed into a 'trellis' by connecting sideways strands, creating a basket-like structure that may help to control the release of spores in very wet places. In North America, the aquatic moss *Scouleria* has dispensed with its peristome altogether, and instead has a little lid that is popped off (but remains attached to a central column) when its round capsules dry out to become doughnut-shaped.

Most aquatic mosses are *rheophytes* – they are found under water for some of the year but are not submerged at other times. There are certain characteristics that these

species have in common, such as leaves that are closely pressed to stems and features such as thickened leaf borders that may be associated with resisting fast-flowing water. Interestingly, it seems that these same features have evolved independently in several unrelated groups of mosses, and that rheophytes tend not to have many close relatives, having been evolutionarily isolated for long periods of time.

> There are a number of cases where mosses that almost always grow on land are found under water and have adopted a completely different growth form.

There are a number of cases where mosses that almost always grow on land are found under water and have adopted a completely different growth form as a result that makes them very hard to identify. One particularly striking example of an aquatic moss habitat is in the Sôya Coast region of East Antarctica, where bizarre 'moss pillars' are found at the bottoms of some ice-free lakes. These column-like structures are made up of two moss species as well as a number of algae and microscopic organisms. While one of the species of moss had been reliably identified as a modified form of a species that is common on the land in Antarctica, the other had confused botanists for years until Kengo Kato and his collaborators used DNA to show that it is a form of *Leptobryum wilsonii*, a species that is commonly found in South America but has never been found growing on the land in Antarctica.

Strange structures known as 'moss pillars' in a
freshwater lake in coastal East Antarctica.
Image: BJ Warnick/Alamy

Temperate Rainforest

In the British Isles, most people are surprised or even sceptical when told that rainforest exists in their own country. We are so accustomed to thinking of rainforest as exotic and tropical, filled with monkeys, toucans and possibly tigers, with colourful frogs on every dripping leaf and a constant rhythmic cacophony of buzzing insects. Yet nearly all of the definitions of rainforest (there have been many!) also encompass the more or less continually humid forests we find in the wetter, western parts of Britain. These forests have gone by many names, including 'Atlantic oakwood' and 'scrub oak', and are now sometimes referred to as 'Celtic rainforest'. Although these forests are now restricted to a few scattered remnants within the already narrow climatic zone that can support them, they are of critical international importance, because this particular type of forest is extremely rare elsewhere in the world.

Temperate rainforest is just as amazing and important as other types of rainforest – not least because perhaps its most recognisable feature is that it is dripping with mosses! Wander into a sheltered oak woodland on the coastal fringes of Argyll in Scotland, or even in parts of Devon or Cornwall, and you will see mosses and liverworts everywhere – covering the ground, blanketing the rocks, and festooning the trunks, branches and twigs of the trees.

Just as with tropical cloud forest, the conditions that support temperate rainforest are a high annual rainfall that is fairly evenly distributed throughout the year (there is no significant dry season), temperatures that don't fluctuate much between the seasons, and overall, temperatures that are relatively cool and not often below freezing. Outside of the tropics these conditions are mostly only found at temperate latitudes in areas strongly

Temperate and boreal rainforest around the world.
Image: Modified by Neil Bell from Mackey et al.
(2017), in turn based on DellaSalla (2011)

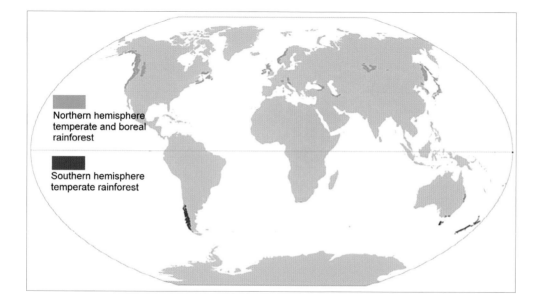

influenced by the sea – so called *oceanic* regions.

Definitions of this type of forest have varied and this has slightly influenced their perceived distribution over the years, but the most unambiguous examples are in Western Scotland, Ireland, Norwegian Fiordland, Western North America, Japan and the adjacent fringes of Eastern China, the western fringes of southern South America, Eastern Australia and Tasmania, and New Zealand. All of these areas have their own slightly different collections of mosses and liverworts, but because the temperate areas of the southern and northern hemispheres are so far from each other and especially because they are separated by the tropics, most of the species found in them are utterly distinct – often they have evolved from completely different families. Within each hemisphere, however, the same or closely related species are frequently found in temperate rainforests on different continents, separated by thousands of miles, while there are a few species that are found in both southern and northern temperate rainforests.

This moss, *Ulota fulva*, is actually growing in tropical cloud forest, although several other very similar-looking species of *Ulota* are found in identical situations in temperate rainforest in Britain. At this scale, tropical cloud forest and British rainforest look very similar! Image: © Des Callaghan

◀▶ The typical 'tree mosses' *Mniodendron comatum* and *Hypopterygium didictyon* in coastal temperate rainforest in New Zealand. Images: © Neil Bell

SOUTHERN TEMPERATE RAINFOREST

From at least 90 million years ago until as recently as 40 million years ago, and despite being situated over the South Pole just as it is now, the continent of Antarctica was covered in lush forest. This seems incredible to us today, but during this period Antarctica was still attached to Australia and to the southern tip of South America (these areas together being the remains of the ancient supercontinent of *Gondwana*), and this made the climate profoundly different. Nowadays, a strong, cool ocean current, sometimes referred to as the *west wind drift*, circles clockwise around Antarctica as it is seen from the South Pole. This isolates it from the warmer currents to the north and makes it much colder than it would be otherwise.

Forty million years ago, however, this current had yet to form because the sea had yet to open up between Antarctica and Australia, and between Antarctica and South America. When this happened, Antarctica rapidly cooled and its forests vanished, although the same forests persisted and continued to evolve in South America, Australia

and the area that would give rise to New Zealand. In Australia, however, although wet forests once existed over much of the continent, they became increasingly restricted to a

◀ The 'pipe cleaner moss' *Ptychomnion aciculare* is very common in temperate rainforest in Eastern Australia and New Zealand. A very similar-looking species, *Ptychomnion cygnisetum*, is abundant in South American temperate rainforest. Image: © Neil Bell

▼ *Weymouthia mollis* in Tasmania, with its typical 'dangly' or pendent habit. Image: © Stephen Thorpe (CC BY)

very narrow eastern coastal fringe as Australia drifted northwards and became progressively more arid. Similarly, in South America they are now only found in the extreme south west, mainly in Chile's Magellanic Region. These wet, oceanic forests are now lush temperate rainforests harbouring an abundance of mosses and liverworts, many from families that are completely unfamiliar to northern botanists and that likely evolved in Gondwana during the last years of the dinosaurs!

Rhizogonium distichum in Australia.
Image: © Neil Bell

The similarities between the rainforest mosses of Eastern Australia and Tasmania, New Zealand and the south-west fringes of South America are striking. In all of these areas species of the large, crinkly-leaved *Ptychomnion* and the swept-over, narrow-leaved *Dicranoloma* are abundant. Large, plush cushions of *Leptostomum* with its cigar-shaped sporophytes occur on trunks and dangly mosses such as *Weymouthia* and *Glyphothecium* festoon the twigs. Dramatic tree mosses and fan mosses (see p. 135)

Large, plush cushions of *Leptostomum* occur on trunks and dangly mosses such as *Weymouthia* and *Glyphothecium* festoon the twigs.

from the families Hypopterygiaceae and Hypnodendraceae are found on the ground and on trees, although the Hypnodendraceae is strangely under-represented in South America. Small mosses with a flattened appearance such as *Rhizogonium* and *Catagonium* are found on dead wood and in shaded hollows. These are rather unusual for mosses in having their leaves more or less in two distinct ranks, very like leafy liverworts.

A few of these wet forest species are the same in all of the southern

Catharomnion ciliatum (left), *Cryptopodium bartramioides* (top right) and *Cladomnion ericoides* (bottom right) are all species that only occur in New Zealand, as well as being the only species in their genera (in other words, *Catharomnion*, *Cryptopodium* and *Cladomnion* only occur in New Zealand). Images: © Neil Bell

temperate rainforests and may disperse by spores quite easily between the southern continents, while in many other cases similar *sister species* are found in South America and Australasia, although the species themselves may only occur in a single country. In the latter case, these may have been isolated for millions of years, having evolved from a common ancestor that once lived in the Antarctic forests of Gondwana.

In all of these regions there are a few uniquely distinct mosses that don't have close relatives anywhere else in the world, even in other southern

Gackstroemia magellanica (left) and a species of *Lepicolea* (right) in temperate rainforest in southern Chile. These robust, regularly branched liverworts look rather moss-like! Images: © Neil Bell

temperate areas, although they are members of distinctly southern temperate families. New Zealand in particular has an especially large number of these (more than ten), including the weird and wonderful *Cladomnion*, *Catharomnion* and *Cryptopodium*. Studies of DNA suggest that many of these unique New Zealand *endemic genera* (unlike seemingly similar ones elsewhere) split off from their closest relatives rather a long time ago, so they may be *paleoendemics* – lineages that evolved in Gondwana and were once found in other places too, but are now restricted to New Zealand.

▶ The giant liverwort *Schistochila appendiculata* in New Zealand. The ferns and young trees in this image give some idea of scale!

▼ *Schistochila appendiculata*

Images: © Neil Bell

Liverworts are also very diverse and abundant in these forests, especially the wettest ones nearest the sea. Quite a few are regularly branched like feathers and look rather like pleurocarpous mosses (see p. 132 and Box), such as species of *Lepicolea* and *Gackstroemia*, while others, such as some species of *Schistochila*, are monsters! New Zealand is particularly rich in liverworts, many of which are found nowhere else. There are about as many species of liverworts as there are mosses, which is highly unusual elsewhere in the world.

The beautiful *Myurium hochstetteri* is known to British botanists as a moss that grows in rock crevices and heathy slopes very near to the sea in the extreme west of Scotland, but in the Azores and Madeira it occurs abundantly in laurisilva forest, a type of temperate, subtropical rainforest. Image: © Des Callaghan

NORTHERN TEMPERATE RAINFOREST

The unique and exotic temperate rainforests of the southern hemisphere have equally spectacular counterparts in the north – and as mentioned above, the British Isles is one of the most important areas for these habitats globally. Truly, in the west of Britain we have woodlands that once seen for what they really are, rival any in the world for abundance and diversity of rare moss and liverwort species. It's only relatively recently that they have been widely recognised as such, however – perhaps reflecting a historical bias against what were once called 'lower' plants – meaning that they were thought to be less evolutionarily advanced (which as we have seen, is not really true), but also perhaps implying that they were somehow less important. More often than not, this simply allowed botanists and ecologists to ignore plants that were seen as too difficult and complicated to identify and survey.

The 'Celtic rainforests' are perhaps seen at their best in steep-sided valleys on the extreme west coast of Scotland, including on the islands. The topography of these places, together with the climatic features described above, helps them to retain the continually moist atmosphere that supports an abundance of rare oceanic moss and liverwort species. This topography has also protected them from human exploitation and grazing, as has their tendency to be found on nutrient-poor soils and uneven, stony ground. In the past they would have occurred more widely within the wettest areas of western Britain, but most rainforest on fertile ground was cleared for agriculture in the prehistoric and early historic periods. Interestingly, the most extensive remaining patches of Celtic rainforest occur in Argyll, which in the early historic period was at the centre of the Gaelic-speaking kingdom of Dál Riata. It seems that the oakwoods of Dál Riata may have been nurtured and protected to some extent because

Typical coastal temperate rainforest in Wales.
This type of 'Celtic rainforest' is found on
the wet, western fringes of Britain and
Ireland where temperatures don't vary
much between summer and winter and
there are a high number of wet days per year.
Image: © Jordan Mansfield/WLTM

of their importance for providing wood for boat building. The Gaelic kingdom had a distinct seafaring culture, being geographically united by the sea more than by land routes.

If you know the mosses and liverworts of temperate rainforest in one hemisphere and then visit the equivalent type of forest at the other end of the world, you experience a strange feeling of the familiar and yet unfamiliar, as if you were in a parallel dimension. The forest has the same structure and overall appearance, with mosses and liverworts everywhere. There are large, crinkly-leaved moss species on the ground and on rocks, while the trunks and branches of the trees are often completely covered with wefts and tight cushions of mosses interwoven with mats of liverworts of all sizes, with other species dangling from the twigs. But the species themselves are mostly very different – indeed, they have truly evolved in

◀ Although very common in a range of habitats, the feathery *Hylocomium splendens* can be particularly abundant on the ground in northern temperate rainforest. Image: © Steve Tuckerman (CC BY-SA)

▼ The crinkly-leaved *Rhytidiadelphus loreus*. Image: © Des Callaghan

parallel to exploit the same living spaces in an alternate but equivalent ecological dimension.

In western British rainforest, many of the species that carpet the ground and the tree trunks are common and are found elsewhere, but are much more abundant and well grown here. Large weft-forming mosses such as the feathery *Hylocomium splendens* and *Loeskeobryum brevirostre* are common on the ground, while the 'mouse-tail moss' *Isothecium myosuroides* is usually ubiquitous on trunks. The large, crinkly-leaved species of *Rhytidiadelphus* and *Hylocomiadelphus* are the northern counterparts of the southern hemisphere *Ptychomnion*, sometimes forming deep spongy carpets. Twigs bristle with little pin-cushions of *Ulota* species, and the large liverwort *Frullania tamarisci* forms rusty-coloured wefts on branches.

But woven among these common plants is a host of generally smaller, rarer species, many of which have their main European populations in Britain and occur in few other places in the world. Tiny liverworts such as *Drepanolejeunea hamatifolia* and *Harpalejeunea molleri* are our equivalents of the species that grow on the evergreen leaves of tropical trees, although here they are found on bark and rock. Several globally uncommon species of the liverwort *Plagiochila* can be abundant, with their beautiful drooping, spiny-toothed leaves. The rare *Pseudomarsupidium decipiens* looks similar in some respects, but has strongly concave leaves, while

The diminutive liverwort *Drepanolejeunea hamatifolia* is in the family Lejeuneaceae, which is most diverse in tropical areas. In Scotland it's found in very wet, sheltered places such as near running water in temperate rainforest. Image: © Des Callaghan

Jubula hutchinsiae has the rear parts of its leaves shaped into tiny flask-like structures (just like *Frullania*, p. 106), but has its main leaf lobes fashioned like miniature holly leaves, complete with large, spiky teeth.

A number of these more unusual species of very oceanic British rainforest have links to tropical regions such as Central America and the Caribbean, where they are found at higher altitudes in cloud forest, and especially to islands in Macaronesia such as the Azores. It's possible that some of them could have evolved from species that were more widespread before the ice ages, when the type of wet forest that is now only found as a few scattered remnants in Macaronesia occurred over large parts of North Africa.

We have focussed here on British temperate rainforest, but as mentioned above we can find equivalent forests in several other oceanic parts of the northern hemisphere. Even with our European bias, however, this discussion would hardly be complete if we didn't

◀ A closeup of the leaves of the liverwort *Pseudomarsupidium decipiens*, a less common species found in very wet rainforest in the extreme west of Britain and Ireland. Like a number of species of our most oceanic forests, it's also found in Macaronesia and in tropical areas such as South America, the Caribbean and tropical Africa. Image: © Des Callaghan

▼ One of the most beautiful and characteristic mosses of North American temperate rainforest, *Dendroalsia abietina* is the only species in its genus and is only found in the Pacific North West. It looks dramatically different when wet (right) and when dry (left). Images: © By Randal

mention the largest extent of coastal temperate rainforest in the world, which stretches all along the Pacific coast of temperate North America, from Northern California in the south to Alaska in the north. This forest looks a little different from our British rainforest, being largely dominated by giant trees such as coastal redwood in California and large conifers such as hemlocks, cedars and Douglas fir in British Columbia. Underneath the canopy, however, mosses and liverworts dominate and often outnumber the species of vascular plants, just as they do in other temperate rainforests. The mosses are often large and impressive, such as the species of *Neckera* that have leaves with deep furrows running across them to hold water, or *Alsia californica* and *Dendroalsia abietina*, each the only species of their kind and only occurring in North America. *Dendroalsia* is a beautiful fan moss that grows on tree trunks, which when dry curls its branches downwards to resemble little clenched fists.

Mountains and Tundra

In very cold parts of the world where sub-zero temperatures persist for most of the year and there is a very short growing season for flowering plants and conifers, it's effectively impossible for forests to exist. The transition point between a climate where native trees can grow and one where they can't is called a *tree line*. On earth, it is mainly geography that determines the temperature of the climate, and the two main factors

Extensive areas at high altitude in temperate regions are very similar ecologically to arctic tundra. The Cairngorm plateau in Scotland is snow-covered for much of the year and many of the plants that grow there (including some mosses) are not found anywhere else in Britain. Such areas are of relatively limited total extent and are potentially threatened by even low-level human activity. Image: © Neil Bell

are *latitude* (i.e. distance from the equator) and *altitude*. The climate generally becomes colder the further away from the equator you get, and the further away from sea level. This means that there are two types of tree lines – the ones you encounter when you go up mountains, and the ones you encounter in Arctic and Antarctic regions.

On the equator, tree lines are usually at altitudes of thousands of metres, while in northern temperate and subarctic regions they are only a few hundred metres. In an abstract sense, the latitudinal tree line begins at latitudes where the altitudinal tree line drops to sea level. In practice it's not that simple, as factors such as shelter, other aspects of the climate and the influence of the ocean affect local tree lines.

The type of treeless tundra you find in the Arctic is very similar to the type of landscape you find on mountain plateaux, and often many of

A typical snowbed habitat in the Cairngorms in Scotland, where small patches of snow may persist almost throughout the year. These habitats are host to some of the rarest mosses and liverworts in the British Isles and are at significant risk from climate change. The snowbed bryophyte communities can be seen as bright green and blackish patches in this picture, especially above the melting upper edge of the snow. Image: © Neil Bell

Snowbeds

In areas where snow and ice are present all year round glaciers form – because the ice never melts completely it is renewed each year, accumulating at the top and then 'flowing' to where it melts at the edges lower down. However, there are some places in the mountains where snow is present for most of the year, but not quite until the next winter snows come. These areas are just short of becoming glaciers, although the snow-free period may be very short. They are often very limited in extent, forming (in Scotland at least) in north and east facing hollows where snow builds up because it is blown there by the prevailing winds, and because it is sheltered from the sun. Any vegetation in these small patches, in the brief part of the year when they are snow-free, is dominated by mosses and liverworts.

In the UK snowbed habitats are home to some of the rarest mosses and liverworts in the country because even when all of these patches are considered together they make up a very small total area, and some of the species that grow there are very particular and aren't found anywhere else. The same species often occur in snowbed habitats in Scandinavia and North America and may also be common in the Arctic, but the British populations are hanging on in these tiny patches and are potentially highly at risk from climate change. If the climate warms and these habitats disappear, the mosses and liverworts that can only grow in these places will have nowhere else to go.

Snowbed mosses and liverworts are often colourful and attractive, but some of the smaller, mat-forming liverworts are truly minute and very hard to identify (see last image on p. 183). Characterising the vegetation of these places and monitoring them for modification due to climate change has been a significant achievement of British moss and liverwort specialists.

Kiaeria starkei, another moss associated with places where snow persists throughout much of the year. It is found in Scotland in snowbed habitats on the highest mountains, but here it is growing in Iceland beside a hillside stream. Image: © Des Callaghan

the same plants can be found growing there. In the UK the Cairngorm plateau is effectively tundra, and indeed some plants (including mosses) have their only UK occurrences in the Cairngorms but are otherwise quite common in the Arctic.

Mosses and liverworts are extremely important in Arctic and Antarctic environments because they are often the dominant plants. They can't grow on glaciers (no plants can), but some species are able to

Some plants (including mosses) have their only UK occurrences in the Cairngorms but are otherwise quite common in the Arctic.

grow in places that are covered in ice and snow for most of the year and where the soil is extremely thin or non-existent, due to being able to survive in a more or less dormant state and to get moisture and nutrients from seepages and from the atmosphere when it is available, rather than through roots. As well as these very cold, mostly snow-covered places in the Arctic and at high altitudes, other types of mountain environments are often dominated by mosses. Even in relatively warm mountainous places, very steep slopes, exposed rock, cliffs and scree are often unable to develop soil, and so species of mosses dominate (just as they do on wall tops in the city) because they are able to grow and thrive directly on rock.

ARCTIC AND ANTARCTIC MOSSES

The further north or south you go, the greater the percentage of the plants that you find tend to be mosses and liverworts, so that by the time you reach the High Arctic, species of mosses and their relatives can outnumber other plants two to one. This is for the reasons outlined previously – not having roots, being able to obtain water and nutrients directly from the atmosphere and melting snow, and being able to tolerate prolonged periods of desiccation and freezing temperatures gives them an advantage over other plants. Plants with roots need to be able to cope with growing on a thin, *active layer* of unfrozen soil that exists more or less periodically on top of the frozen *permafrost*, and rather few of them are able to do this very well. Thus, many places in the Arctic are covered in low-growing cushions of mosses with little in the way of other vegetation.

Because of the dominance of mosses in many Arctic and Antarctic areas, they are a significant food source (together with lichens) for animals such as deer, birds and small mammals. This is unusual in other parts of the world because mosses have a rather low nutrient content for their size (they are largely water), and so are rarely worth the effort of eating when other plants are available. In the Arctic there is little choice, however!

There are many species of Arctic and Antarctic mosses (perhaps around 1,000), but most of them belong to quite a small number of families. Because of this, and because they are highly restricted in their growth forms by the environment, they can be quite hard to tell apart. It's possible that more species exist than we know about, because species that are actually distinct may look practically identical. Also, there are many areas of the

While some of the rare mosses of snowbed communities are fairly large and attractive (top left: *Polytrichastrum sexangulare*, top right: *Kiaeria falcata*, bottom left: *Conostomum tetragonum*), many of the important liverworts are truly tiny and very hard to identify (e.g. bottom right: *Marsupella brevissima*), especially in situ in the sort of weather conditions typical of high mountains in Scotland! Images: © David Genney

Arctic and Antarctic that haven't been explored for mosses because they are difficult to get to. A good number of Arctic mosses have distributions that are described as 'Circumpolar Boreo-arctic montane', meaning that they are found in a range of cold environments in the northern hemisphere, including high mountains in temperate areas.

The colourful *Bryum weigelii* is found in wet springs and flushes in the Arctic, as well as in similar places in the mountains in temperate areas. Image: © Des Callaghan

1,500-YEAR-OLD FROZEN MOSSES COME BACK TO LIFE!

Across the Arctic, glaciers are retreating due to warming of the climate, and this has been especially marked in the last 20 years. In the Canadian Arctic it was noticed that some mosses that had been entombed in the ice had started to grow again once thawed, and their ability to do this was confirmed in laboratory experiments. This is especially remarkable, because radiocarbon dating showed that these plants had been frozen into the glacial ice at the beginning of the period known as the 'little ice age' when the glaciers expanded, about 400 years ago. Even more amazingly, subsequent to that discovery mosses from Antarctica that were entombed in ice 1,500 years ago have also been revived in laboratory conditions!

Of course, the old stems and branches of these plants don't simply come alive again after 400 years of being frozen, but parts of them are able to start growing once more, producing new growth that is part of the same individual. This shows how mosses are able to survive extreme conditions through a combination of staying alive (although dormant) in such conditions and the process called *totipotency* (see p. 31). This has important implications for how we think about mosses surviving periods of ice advance and retreat, as has happened repeatedly throughout successive ice ages. It used to be assumed that when glaciers retreated, mosses and other plants would all need to recolonise the area from elsewhere. But it may be that some moss species are able to survive ice ages in the same place, as fragments frozen in glaciers.

MOUNTAIN MOSSES

Above the tree line, where the mountain forests can't exist, we find true *montane* habitats, dominated by grasses and small shrubs, as well as mosses and their relatives.

Places above the tree line that are dominated by shrubs such as heather and blueberry or by coarse, tufted grasses are called *heaths*. Heathland can also occur below the natural tree line if tree growth is prevented in some way, such as by artificially high levels of grazing (as is the case in much of the British uplands). Even when heaths are completely dominated by small shrubs, mosses can usually also be found growing in abundance as a miniature understorey underneath the shrub layer. Wetter heaths in particular can often harbour a diverse collection of mosses and liverworts, and there is a special, very interesting type of mountain heath that is only found in the wettest regions of the world near to oceans and is dominated by a collection of highly unusual, large liverworts (see 'Oceanic montane heath' below).

As the climate becomes colder at higher altitudes still, the soil becomes thinner, and the ground is often frozen or covered in ice and snow for much of the year. Grasses and shrubs such as heather become much more restricted and mosses and their relatives often dominate, just as they do in the High Arctic. There is a particular collection of specialised mosses and liverworts associated with the areas where the snow lasts longest into the summer (see Box, p. 179).

Steep slopes and erosion in mountains often create a familiar landscape of exposed rock, boulders and cliffs that most plants are unable

Encalypta ciliata, a moss that in northern temperate regions is found in crevices and on rock ledges in the mountains. Image: © Des Callaghan

to grow on, even below the natural tree line. Mosses and their relatives can be extremely diverse in such places, due to the very wide range of different types of *microhabitat* created by combinations of different types of rock and differences in moisture availability and shade. Limestone and other types of rock that have an alkaline chemistry will have a completely different collection of moss species growing on them compared to ones found on acidic rock such as granite or sandstone. Little ledges and crevices, sheltered areas at the bases of cliffs and areas of scree can all provide homes for specialised mountain mosses and liverworts, some of which can be very rare because they have very exacting requirements for precise combinations of climatic and chemical factors.

OCEANIC MONTANE HEATH

At moderate altitudes in the wettest parts of Britain, Ireland and South West Norway a very particular type of liverwort-dominated heath is found in the dampest and most sheltered parts of the hills. These places are often in what are called *corries* in Scotland (*cwms* in Wales) – hollows formed by glaciers in the sides of mountains – although they may just be steep slopes in other sheltered areas. They are nearly always north or east facing, ensuring that they hardly ever experience direct sunlight and are protected from the prevailing winds, and they are always found in places that have a very *oceanic* climate – one that is continually wet throughout the year and with temperatures that don't vary very much between summer and winter. In the British Isles the co-occurrence of mountains

The liverwort *Adelanthus lindenbergianus*, with the 'swept over' look that many of the robust, stringy liverworts of the oceanic montane heath community have. This particular species is very rare in the wettest parts of Scotland and Ireland, otherwise being found only in Central and South America, tropical and Southern Africa, and a few islands in the southern hemisphere. New information from DNA suggests that the populations in these areas may in fact represent more than one distinct species. Image: © Neil Bell

with such a climate occurs in North West Scotland, Western Ireland and to a lesser extent North Wales and the Lake District.

The liverworts that grow in these places form tall, stringy cushions scattered among low heather, scree and more common mosses and liverworts. They are unusually robust for liverworts that are growing in the open, with many species having a 'swept over' appearance, with the two main rows of leaves folded in on each other to retain moisture.

What is truly remarkable about these plants, however, is their global distributions. Several species occur mainly in Britain and Ireland, Norway, Western North America and the Himalayas, while a few are practically only found in the British Isles and the Himalayas. *Plagiochila carringtonii*, for example, is only known from Britain and Ireland,

▼ *Plagiochila carringtonii*, another oceanic montane liverwort with a very restricted and puzzling global distribution (in the picture on the right it's the pale yellow plants growing together with the red-orange *Pleurozia purpurea*, a more common species). Outside Scotland, Ireland and the Faroe Islands it's only found across the Himalayas and in the adjacent mountains of Yunnan, China! Images: © Neil Bell

▶ *Herbertus borealis* is found in Scotland, and nowhere else in the world! Image: © Neil Bell

the Faroe Islands and across the wider Himalayas (including Yunnan in China), while *Anastrophyllum alpinum* occurs only in Scotland, the Himalayas and one of the Aleutian Islands. It's likely that these species have dispersed from the Himalayas to Britain at some point since the last ice age (via spores or perhaps tiny leaf fragments) and that they are found nowhere else due to a combination of the extreme global rarity of the places they are able to grow in, and the relatively low chances of dispersal.

One species, *Herbertus borealis*, appears to occur nowhere else in the world outside of Scotland. It's possible that it may yet be discovered in the Himalayas, although for the time being it is perhaps Britain's most

unique and interesting liverwort (as well as one of the most attractive!).

Although these plants are described here rather than in either of the sections dealing with forest, they are arguably not true mountain plants. In the Himalayas they are found on the ground just at or below the tree line, where the last hardy trees hang on and form a very short, open canopy. It may be that when Scotland still had its natural tree line they were found in similar situations and that they are now mainly restricted to a type of surrogate mountain heath habitat. Indeed, there are one or two places in Scotland where the natural tree line still exists and these liverworts are found on rough ground under stunted trees.

Bryum capillare is a common species on urban wall tops, as well as in natural environments such as this. Image: © David T. Hollyoak

Disturbed and Urban Environments

If you go for a walk in the city on a sunny spring day just after rain and look closely at the tops of the walls, you might find miniature gardens of mosses stretching for miles. On newer brickwork walls these may form strips, like tiny herbaceous borders, along the lines of the mortar surrounding the bricks. On older brick or stone walls they may cover the entire upper surface and some of the sides of the walls too.

Larger, creeping moss may be the lime-loving pleurocarpous species *Homalothecium sericeum*, which when dry tightly curls the short branches of its long golden shoots up and over itself. Many other mosses form dense, rounded cushions around 1–4 centimetres across with multiple 'blobs-on-sticks' (sporophytes) protruding from them.

If you look more closely still, you will see that these little pin-cushions are not all the same. There will likely be the hoary, grey-green cushions of *Grimmia pulvinata* with its long white hairs and sporophytes with arched necks and heads buried in the leaves. Also with white hairs but with greener, shorter and wider leaves, *Tortula muralis* will have its sporophytes sticking straight up with long, pointed tips. And there may be *Bryum capillare*, with larger, shinier leaves and beautiful large fat, drooping sporophytes resembling ripe fruit. Often cushions of all of these species will be growing tightly pressed together, fighting for precious space on the lime-rich mortar or brick.

These particular mosses (and others) belong to a group of largely unrelated species that are all adapted to lime-rich surfaces, can tolerate relatively high levels of nitrogen in the air and are able to withstand frequent drying out in open, sunny places. Many of the features they have in common are adaptations to these conditions – for example, growing in tight cushions reduces moisture loss, while hairs on the tips of leaves can act as focal points for

condensation of water at night. As a result they are at home in our cities, and because they usually produce many spores and are used to moving around, they can exploit the often-temporary nature of urban environments.

Waste ground and undisturbed concrete surfaces, such as in abandoned demolition sites and quarries, can be rather rich in moss species, a few of which can be quite unusual. The actual plants which occur will depend on the chemistry of both the waste (e.g. concrete) and the soil, but often present will be extensive mats of pointed-leaved, often untidy-looking *Didymodon* species, as well as *Ceratodon purpureus* with its abundant purple-stalked sporophytes. The large pleurocarp *Brachythecium rutabulum* may be found in these places, despite

Growing in tight cushions reduces moisture loss, while hairs on the tips of leaves can act as focal points for condensation of water at night.

also being one of the commonest mosses of stones and wood on the ground in forests. Even on heavily-used pavements in the middle of cities, tiny mosses such as the silver-tipped *Bryum argenteum* can be found growing in the cracks between the paving slabs.

Mosses and their relatives don't grow as abundantly on trees in city centres as they do in more natural and less polluted places, but sometimes little stems of star-shaped *Zygodon* species form a patchy green crust on shaded trunks, and the short-hair-pointed *Orthotrichum diaphanum* with its sporophytes hidden among its leaves occurs as little tufts. Any untidy-looking, creeping or feathery-looking moss on a more shaded branch is fairly likely to be the very common woodland species *Hypnum cupressiforme*.

RARE MOSSES OF MAN-MADE HABITATS

We're used to hearing of plants and animals being driven to extinction by human activities, and indeed we are now in the midst of the latest of several mass-extinction events in earth's history. This time, however, rather than disaster-movie-style catastrophes such as meteorite impacts or the opening of giant volcanic fissures in the earth's crust, it is our own exploitation of the planet's natural resources that's responsible, and particularly the ongoing catastrophic destruction of natural habitats such as old-growth forests and wetlands. Mosses are just as much victims of these activities as any other plant or animal, and the best way to conserve threatened mosses is to conserve the threatened habitats they are part of, together with all of the other forms of life they support.

Given this depressing background it's refreshing to hear of tiny exceptions to the rule – in this case, situations where human activities have helped to create space for rare mosses.

The heavy-metal mosses are a small collection of very specialised species that are able to grow on soil that is highly toxic to most other life, such as in the spoil from lead, copper and zinc mining activities. The lead moss *Ditrichum plumbicola* was only discovered in 1976 in Caernarvon in Wales, and was once thought to occur nowhere outside of Britain (it's now also known from a small number of sites in South West Germany and Belgium). It occurs at a very few old mining sites in the UK, some of which are ancient spoil heaps from Roman and medieval mining that left a lot of lead still in the soil that was subsequently re-processed in the 19th century. Similarly, the rare 'copper moss' *Scopelophila cataractae* is found in waste heaps surrounding old mine sites in Wales and Cornwall (often zinc rather than copper mines), and other places around the world that have heavy-metal-rich soils.

With the cessation of mining activities at most of these British sites, however, the heavy-metal soils that these species grow on are no

The heavy-metal mosses are a small collection of very specialised species that are able to grow on soil that is highly toxic to most other life.

longer being continually recreated, and gradual leaching of metals by rain will slowly make the soil suitable for other plants to grow in. These other species will likely outcompete the rare heavy-metal mosses, which are present largely because they are the only plants that are currently able to grow there. It may be that the best way to conserve these species is to create new habitat for them by periodic exposure of the metal-rich soils through deliberate disturbance. It seems that these mosses are able to tolerate such extreme conditions because they can get much of the water they need from the atmosphere rather than the soil, and because they are able to minimise the quantities of heavy-metals that enter their cells (storing them on their external surfaces instead), as well as to tolerate relatively high concentrations inside their tissues.

Another example of a moss that in the past has been common on mine waste is the so-called 'bug moss', *Buxbaumia aphylla* (see Box, p. 202). This evolutionarily fascinating and rather rare species seems to be very fussy about the particular type of disturbed, sandy soil it grows on. Thirty years ago it was quite common on the 'oil-shale bings' of West Lothian in Scotland. These are giant artificial hills made of debris from shale mining undertaken in the late-19th to mid-20th century. It seems that the 'bug moss' was only happy there during a particular phase of the development of these sites from raw waste to fully vegetated ground. Now that grasses and small trees have taken over at these sites, it is no longer found there.

The very rare moss *Ditrichum plumbicola* is only found at a few old mining sites where the soil is heavily contaminated with lead. It may rely, to some extent, on the continuous creation of new habitat through disturbance.
Image: © Des Callaghan

202 | Moss habitats

Buxbaumia – the 'bug moss'

Buxbaumia is one of the strangest of all mosses – not least because in one important way, it doesn't live like other mosses do. It has been mentioned how mosses have two quite different phases to their lifecycles, each of which is, in a sense, a distinct individual (see p. 28). The green, leafy *gametophyte* is responsible for making food using energy from the sun (*photosynthesis*) and for sexual reproduction, which in turn creates the *sporophyte* – the non-leafy 'blob-on-a-stick' that produces and releases spores.

In *Buxbaumia*, however, the gametophyte is nothing more than microscopic filaments and is effectively leafless (except for one or two tiny leaves that protect the sexual structures). The sporophyte, however, is massive in comparison and a rather unusual flattened, asymmetrical shape (see opposite), this being responsible for its many nicknames, including 'bug moss' and 'bug-on-a-stick'! As the sporophyte is green while it is growing and must need energy, it is likely that it is significantly photosynthetic, although it's possible that the filamentous gametophyte may provide some resources and that the plant may have associations with soil fungi that provide food (although there is little evidence for this).

There are 12 species of *Buxbaumia* worldwide. Its position on the evolutionary tree of mosses fits with how unusual it is. *Buxbaumia* belongs with the mosses that have an 'active' type of toothed structure (the peristome) to control the release of the spores. However, it branches off from this group before all of the other members do – the tree of life splitting at this point to produce *Buxbaumia* on one branch and all of the other 'active toothed' mosses (i.e. the large majority of all mosses) on the other. Its toothed peristome is also unlike any found in the other 'active toothed' mosses and may provide clues about how this important structure evolved.

Species of *Buxbaumia* are generally rare and one of the two British species is protected by legislation, along with a small number of other mosses and liverworts. Most of the other protected species of moss, although just as rare or more so, are fairly closely related to many other species that are quite similar in their form and lifestyle. Sometimes botanists debate whether 'unusualness' should be taken into account when deciding how much protection to offer to rare species. If resources are limited, should all rare and threatened species receive the same protection, or should more unusual ones count for more because by themselves they represent a greater proportion of the evolutionary tree and of total diversity? What do you think?

Spore capsules of the 'bug moss', *Buxbaumia aphylla*.
Image: By Shaun Pogacnik (Public Domain)

ARABLE MOSSES, LIVERWORTS AND HORNWORTS

In the UK, and almost certainly in other countries too, arable fields (fields used to grow crops) are host to a surprisingly large number of species of mosses and their relatives. You might think that because this type of land is dominated by intensively grown single crop species in the summer and looks pretty barren in the winter, there wouldn't be much opportunity for tiny plants such as mosses to gain a foothold. However, many mosses and liverworts are very small plants with very fast lifecycles and have evolved specifically to be able to exploit short-term habitats before larger, more lackadaisical plants can get a foothold.

These mosses spring up quickly from spores and from other, non-sexually produced structures that persist in the soil (*gemmae, bulbils and rhizoidal tubers*), primed and ready to grow as soon as the conditions are right. The green plants mature rapidly over a few weeks or months and produce *sporophytes* and then more spores, completing their lifecycle before the ground is dominated again by crops.

Nearly 100 species of mosses and their relatives are known from arable land, almost 10% of the total number of species known in Britain. Of course, most of these species also grow in other places, but quite a few don't, and are absent or extremely rare outside arable fields. Some of these are rather uncharismatic and might seem almost dull to anyone but a keen bryologist, but others are truly weird as well as rare. The liverwort *Sphaerocarpos*, for example, commonly known as the balloonwort, is one of the 'complex' thalloid liverworts we mentioned

◀ Species of *Riccia* are often found in disturbed habitats, such as on paths and at the margins of arable fields. This is *Riccia cavernosa*, sometimes found in damp places in arable fields and gravel pits. It is a type of complex thalloid liverwort. Image: © Des Callaghan

▼ The balloonwort, *Sphaerocarpos* (a type of liverwort), has its sexual organs hidden in highly unusual and distinctive spherical structures. It is one of the tiny gems that can sometimes be found at the margins of arable fields in Britain at the right time of year. There are two species in Britain, but they can only be told apart by examining their spores. Image: By John Kees (Public Domain)

(see p. 69), but its tiny thallus is largely obscured by a mass of green, hollow spheres that protect the sexual structures and then the sporophytes.

Finally, the four hornwort species that occur in Britain are all found most commonly in arable habitats. Remember, hornworts are an evolutionarily distinct group in their own right, just as mosses and liverworts are. We haven't talked about them much in this book because there aren't many species, but it's strange to think that one of the best places to look for representatives of this unique and long-isolated branch of the tree of life is in a muddy field in Britain in November!

It's thought that mosses in arable fields are becoming less common in Britain, and this is likely due to changes in agricultural practices. In the past, after fields were harvested, they were left as stubble until late October, allowing some arable moss species to more easily complete their lifecycle. It is now more common for fields to be ploughed and sown again just a few weeks after harvesting, potentially interrupting the growth of these species. Increased use of herbicides and fertilisers may also have had a negative impact on arable mosses and their relatives.

The moss *Paludella squarrosa*, a mostly boreal and arctic species that was previously found at a few sites in Great Britain but is now extinct there. It is known from one remaining site in Ireland. Image: © Des Callaghan

Final thoughts

If this book has a single aim, it is to provide its readers with a signposted route into the usually hidden world of mosses. Nearly everyone can appreciate the intricate beauty of an orchid or a spectacular display of cherry blossom, but very few may have noticed the comparable beauty of a moss peristome or a *Sphagnum* leaf. Mosses are hidden by their scale and, although we may walk past them every day, a learned indifference can prevent us from seeing them.

An increased awareness of mosses will be important to ensure their conservation in the face of growing threats. Because mosses are so closely linked to the places where they naturally grow (it's rather difficult to cultivate individual species of moss in gardens or glasshouses), protecting them is largely a case of protecting these places, along with all of the other plants and animals that occur there. Some of these areas are already widely recognised as being important for wildlife conservation, but others are not.

Globally, habitat destruction and modification are the most significant threat to mosses and their relatives, followed by climate change and its consequences (direct and indirect). In addition to the well-known and rapidly ongoing destruction of natural forests, the increased use of water for agriculture and energy, including dam construction and other modifications to natural aquatic systems, is highly detrimental to many bryophyte habitats. Climate warming is making wildfires and droughts more frequent and extreme. Forest wildfires, in southern Europe for example, can devastate moss communities, but fires are also becoming more frequent in other habitats such as heathlands in Britain (sometimes started deliberately by so-called muirburn, a practice used in areas managed for grouse shooting). Droughts that may have been survivable by mosses and liverworts when they were rare and short-lived may not be when they are frequent and prolonged – a concern for some of the internationally important bryophyte habitats in the west of Britain and Ireland. The frequency of such extreme events, combined with other changes associated with climate warming, are changing the maps of where individual moss species are able to grow and survive long-term.

In some cases these changes will mean that species currently growing in one area may theoretically be able to grow somewhere else in the future, but they may not ever be able to get there, as the rate of climate change may outpace the rate at which species can move and become established. In many other cases species will have nowhere to go – this is particularly true for rare mountain communities such as the snowbed mosses and liverworts in Scotland.

Moss-rich habitats are often threatened in unexpected ways. For example, in the British Isles one of the biggest threats to bryophyte-rich temperate rainforest is the non-native common rhododendron (*Rhododendron ponticum*), an invasive species that shades out mosses and other plants and is extremely difficult to control. Pollution also strongly impacts mosses, and while this, not surprisingly, includes air pollution, it also encompasses less well-known agricultural pollution including nitrogen enrichment from agriculture and grazing. Tourism and recreation may also cause problems, including poorly regulated development in natural coastal and alpine habitats and even trampling and erosion on mountain summits and in other remote areas of limited total extent.

The advent of digital photography has made the world of mosses and their relatives more approachable, as evidenced in particular by Des Callaghan's wonderful images in this book. Macro photography of mosses on old-fashioned film was a very hit-and-miss affair – modern digital cameras make the process much easier (if never easy), as shown by the dramatic increase in the number and quality of photographs of mosses in the last 10 or 15 years. As these images become more discoverable online and more widely distributed in books, I hope that mosses will become less obscure and that more people will be curious about them. Who knows – perhaps in a few years' time, virtual reality technology will allow us to scramble through the moss thickets with the mites and tardigrades!

Ultimately, the best way to protect mosses will be to make them more visible and better known. I hope that this book will go a small way towards doing that!

References

This book describes general aspects of bryophyte biology that have been discovered through the work of hundreds of scientists over many years. However, the results of a number of specific research studies have been mentioned throughout the text and the corresponding publications are listed below.

P. 23

GAO B., LI X., ZHANG D., LIANG Y., YANG H., CHEN M., ZHANG Y., ZHANG J. & WOOD A.J. (2017). 'Desiccation tolerance in bryophytes: the dehydration and rehydration transcriptomes in the desiccation-tolerant bryophyte *Bryum argenteum*.' *Scientific Reports* 7: article no.7571.

OLIVER M.J., VELTEN J. & MISHLER B.D. (2005). 'Desiccation tolerance in bryophytes: a reflection of the primitive strategy for plant survival in dehydrating habitats?' *Integrative and Comparative Biology* 45: 788–799.

P. 34, PP. 46–47

COX C.J., LI B., FOSTER P.G., EMBLEY T.M. & CIVÁ P. (2014). 'Conflicting phylogenies for early land plants are caused by composition biases among synonymous substitutions.' *Systematic Biology* 63: 272–279.

HARRIS B.J., CLARK J.W., SCHREMPF D., SZÖLL SI G.J., DONOGHUE P. C.J., HETHERINGTON A.M. & WILLIAMS T.A. (2022). 'Divergent evolutionary trajectories of bryophytes and tracheophytes from a complex common ancestor of land plants.' *Nature Ecology & Evolution* 6: 1634–1643.

PUTTICK M.N. *ET AL.* (2018). 'The interrelationships of land plants and the nature of the ancestral embryophyte.' *Current Biology* 28: 733–745.

SOUSA F., FOSTER P.G., DONOGHUE P.C.J., SCHNEIDER H. & COX C.J. (2018). 'Nuclear protein phylogenies support the monophyly of the three bryophyte groups (Bryophyta Schimp .).' *New Phytologist* 222: 565–575.

SU D., YANG L., SHI X., MA X., ZHOU X., HEDGES S.B. & ZHONG B. (2021). 'Large-scale phylogenomic analyses reveal the monophyly of bryophytes and Neoproterozoic origin of land plants.' *Molecular Biology and Evolution* 38: 3332–3344.

WICKETT N.J. *ET AL.* (2014). 'Phylotranscriptomic analysis of the origin and early diversification of land plants.' *PNAS* 111: E4859-E4868.

P. 48

TOMESCU A.M.F., BOMFLEUR B., BIPPUS A.C. & SAVORETTI A. (2018). 'Why are bryophytes so rare in the fossil record? A spotlight on taphonomy and fossil preservation.' pp. 375–416. In: KRINGS M., HARPER C.J., CÚNEO N.R. & ROTHWELL G.W. (eds.), *Transformative Paleobotany: Commemorating the Life and Legacy of Thomas N. Taylor*. Elsevier Inc.: Academic Press, London.

P. 51

LIU Y. *ET AL.* (2019). 'Resolution of the ordinal phylogeny of mosses using targeted exons from organellar and nuclear genomes.' *Nature Communications* 10: 1485.

P. 54

DUCKETT J.G., PRESSEL S., P'NG K.M.Y. & RENZAGLIA K.S. (2009). 'Exploding a myth: the capsule dehiscence mechanism and the function of pseudostomata in *Sphagnum*.' *New Phytologist* 183: 1053–1063.

P. 65

FORREST L.L., DAVIS E.C., LONG D.G., CRANDALL-STOTLER B.J., CLARK A. & HOLLINGSWORTH M.L. (2006). 'Unraveling the evolutionary history of the liverworts (Marchantiophyta): multiple taxa, genomes and analyses.' *The Bryologist* 109: 303–334.

P. 66

RENZAGLIA K.S., SCHUETTE S., DUFF R.J., LIGRONE R., SHAW A.J., MISHLER B.D. & DUCKETT J.G. (2007). 'Bryophyte phylogeny: advancing the molecular and morphological frontiers.' *The Bryologist* 110: 179–213.

P. 97

PESCHEL K., NORTON R.A., SCHEU S. & MARAUN M. (2006). 'Do oribatid mites live in enemy-free space? Evidence from feeding experiments with the predatory mite Pergamasus septentrionalis.' *Soil Biology & Biochemistry* 38: 2985–2989.

P. 102

WESOŁOWSKI T. & WIERZCHOLSKA S. (2018). 'Tits as bryologists: patterns of bryophyte use in nests of three species cohabiting a primeval forest.' *Journal of Ornithology* 159: 733–745.

P. 105

MANNO P., RAGUSO R. & GOFFINET, B. (2009). 'The ecology and evolution of fly dispersed dung mosses (Family Splachnaceae): manipulating insect behaviour through odour and visual cues.' *Symbiosis* 47: 61–76.

P. 107

BARTHLOTT W., FISCHER E., FRAHM J.-P. & SEINE R. (2000). 'First experimental evidence for zoophagy in the hepatic Colura.' Plant Biology, 2: 93–97.

HESS S., FRAHM J.-P. & THEISEN I. (2005). 'Evidence of zoophagy in a second liverwort species, Pleurozia purpurea.' *The Bryologist* 108: 212–218.

P. 118, P. 123

GLIME J.M. (2019). 'Tropics: cloud forests, subalpine, and alpine.' Chapt. 8–10. In: GLIME J.M., Bryophyte Ecology. Volume 4: Habitat and Role. Ebook sponsored by Michigan Technological University and the International Association of Bryologists.

PP. 125–126

BRODRIBB T.J., CARRIQUÍ M., DELZON S., MCADAM S.A.M. & HOLBROOK N.M. (2020). 'Advanced vascular function discovered in a widespread moss.' *Nature Plants* 6: 273–279.

P. 128

FELDBERG K., SCHNEIDER H., STADLER T., SCHÄFER-VERWIMP A., SCHMIDT A.R. & HEINRICHS J. (2014). 'Epiphytic leafy liverworts diversified in angiosperm-dominated forests.' *Scientific Reports* 4: article no. 5974.

P. 136

BELL N.E., NEWTON A.E. & HYVÖNEN J. (2012). 'Epiphytism and generic endemism in the Hypnodendrales: *Cyrtopodendron*, *Franciella* and macro-morphological plasticity.' *TAXON* 61: 498–514.

P. 147

MIRONOV V.L ., KONDRATEV A.Y. & MIRONOVA A.V. (2019). 'Growth of *Sphagnum* is strongly rhythmic: contribution of the seasonal, circalunar and third components.' *Physiologia Plantarum* 168: 765–776.

P. 150

KATO K., ARIKAWA T., IMURA S. & KANDA H. (2013). 'Molecular identification and phylogeny of an aquatic moss species in Antarctic lakes.' *Polar Biology* 36: 1557–1568.

MISHLER B.D. & SHEVOCK J.R. (2014). [Abstract] 'The evolution and ecology of rheophytic mosses.' p. 36. In: *Botany 2014: Frontiers in Botany*. Abstract Book. Botanical Society of America, Boise, ID.

NAKAI R., ABE T., BABA T., IMURA S., KAGOSHIMA H., KANDA H., KOHARA Y., KOI A., NIKI H., YANAGIHARA K. & NAGANUMA T. (2012). 'Eukaryotic phylotypes in aquatic moss pillars Inhabiting a freshwater lake in East Antarctica, based on 18S rRNA gene analysis.' *Polar Biology* 35: 1495–1504.

P. 152, P. 167

ELLIS C.J. (2016). 'Oceanic and temperate rainforest climates and their epiphyte indicators in Britain.' *Ecological Indicators* 70: 125–133.

P. 153

DELLASALA D.A. (2011). *Temperate and Boreal Rainforests of the World: Ecology and Conservation*. Island Press, Washington.

MACKEY B., CADMAN S., ROGERS N. & HUGH S. (2017). 'Assessing the risk to the conservation status of temperate rainforest from exposure to mining, commercial logging, and climate change: a Tasmanian case study.' *Biological Conservation* 215: 19–29.

P. 163

BELL N.E, PEDERSEN N. & NEWTON A.E. (2007). '*Ombronesus stuvensis*, a new genus and species of the Ptychomniaceae (Bryophyta) from south west Chile.' *TAXON* 56: 887–896.

BELL N.E, QUANDT D., O'BRIEN T. J. & NEWTON A.E. (2007). 'Taxonomy and phylogeny in the earliest diverging pleurocarps: square holes and bifurcating pegs.' *The Bryologist* 110: 533–560.

SHAW A.J., HOLZ I., COX C.J. & GOFFINET B. (2008). 'Phylogeny, character evolution, and biogeography of the Gondwanic moss family Hypopterygiaceae (Bryophyta).' *Systematic Botany* 33: 21–30.

PP. 167–170

BAIN C. (2015). *The Rainforests of Britain and Ireland: A Traveller's Guide.* Sandstone Press, Dingwall.

P. 179

FLAGMEIER M., LONG D.G., GENNEY D.R., HOLLINGSWORTH P.M., ROSS LC. & WOODIN S.J. (2013). 'Fifty years of vegetation change in oceanic-montane liverwort-rich heath in Scotland.' *Plant Ecology & Diversity* 7: 457–470.

ROTHERO G., GRYTNES J.-A., BIRKS H. J. B. & GENNEY D. (2007). *Effects of Climate Change on Bryophyte-Dominated Snowbed Vegetation. Part 1: Cairngorms.* Scottish Natural Heritage: Interim Report, contract number B201716.

P. 186

LA FARGE C., WILLIAMS K.H. & ENGLAND J.H. (2013). 'Regeneration of little ice age bryophytes emerging from a polar glacier with implications of totipotency in extreme environments.' *PNAS* 110: 9839–9844.

ROADS E., LONGTON R.E. & CONVEY P. (2014). 'Millennial timescale regeneration in a moss from Antarctica.' *Current Biology* 6: PR222–R223.

P. 192

BELL D., LONG D.G., FORREST A.D., HOLLINGSWORTH M.L., BLOM H.H. & HOLLINGSWORTH P.M. (2012). 'DNA barcoding European *Herbertus* (Marchantiopsida, Herbertaceae) and the discovery and description of a new species.' *Molecular Ecology Resources* 12: 36–47.

FLAGMEIER M., SQUIRRELL J., WOODHEAD M., LONG D.G., BELL N.E., RUSSELL J., POWELL W. & HOLLINGSWORTH P.M. (2020). 'Globally rare oceanic-montane liverworts with disjunct distributions: evidence for long-distance dispersal.' *Biodiversity and Conservation* 29: 3245–3264.

P. 199

CALLAGHAN D.A. & SAMSON L. (2022). 'Population status and ecology of the globally threatened moss *Ditrichum plumbicola* Crundw. on the Isle of Man.' *Journal of Bryology* 44.

STANKOVI J.D., SABOVLJEVI A.D. & SABOVLJEVI M.S. (2018). 'Bryophytes and heavy metals: a review.' *Acta Botanica Croatica* 77: 109–118.

Index

References to photographs are in *italics* with the page number for the caption in brackets if not on the same page. References to diagrams are indicated in **bold**.

A

Africa

liverworts

Adelanthus lindenbergianus *190–191*

Pseudomarsupidium decipiens *174* (175)

sub-Saharan mountains, tropical cloud forests 123

agricultural practices 167, 205, 208, 209

Aleutian Islawnds, *Anastrophyllum alpinum* (liverwort) 192

algae 30, 42, 44–45, 142, 150

filamentous (thread-like) *green algae* 42

alternating generations 31, 44

altitude 177

amber, fossils in 49

animals

arthropods 99

beetles 91, 97

birds 102, *103* (102)

coneheads 91, 97

dung flies 104–105, *104*

insects 91, 106, 146

spore dispersal by 102, 104–105, *104*

mites 91, 94, *94–95*, 97

moss habitats

for larger animals 102

Sphagnum bogs 146

for tiny animals 90–91

rotifers 91, 97, 100, *101*

spiders 91, 97

springtails 91, 94, *96–97*, 97

tardigrades ('water bears') 91, *98*, 99

worms 91, 94, 97

Antarctica

forests of Gondwana 157, 159, 162, 163

moss pillars (Sôya Coast region) 150, *151*

mosses and liverworts 115, 182–183, 186

tundra 177, 180

aquatic bryophytes 148, 150

arable bryophytes 204–205

Arctic region

mosses and liverworts 115, 180, 182–183, 186

Bryum weigelii (moss) *184–185*

snowbeds 179

tundra 177, 180

arthropods 99

Atlantic oakwood 152

Australasia 8, 126, 162

Australia

mosses

Ptychomnion aciculare ('pipe cleaner moss') *158* (159)

Rhizogonium distichum *160* (161)

temperate rainforests 153, 157, 159, 161, 162

tropical montane cloud forests (Eastern Queensland) 123

Azores

mosses

Myurium hochstetteri 166 (167)

Pseudomalia webbiana 4–5

tropical cloud forests 173

B

bacteria 88, 100, 107

blue-green bacteria 79

Balfour, Isaac Bayley 88

beetles 91, 97

Belgium, *Ditrichum plumbicola* (moss) 198, *200–201*

Bell, Neil, professional background 7–8

birds

nesting material 102

wren with moss *103* (102)

Borneo, tropical cloud forests 123

Botanical Museum, Helsinki, Finland 8

Britain

hornworts, arable hornworts 204–205

liverworts

arable liverworts 204–205

Plagiochila 191

Pseudomarsupidium decipiens 174 (175)

Sphaerocarpos (balloonwort) 204–205, *205*

mosses

arable mosses 204–205

Ditrichum plumbicola 198, *200–201*

Fontinalis 148, *148*

Paludella squarrosa 206–207

pleurocarps 133

Ptychomitrium polyphyllum 110–111

Ulota 154–155

muirburn on heathlands 208

oceanic montane heath 189, 191–192

rainforests 152, 167, *168–169*, 171, 173

snowbeds 179

see also Scotland; Wales

British Bryological Society 9

Brodribb, Timothy 125–126

bryophytes 14–15, 34, 46, 47, 48, 50, 69

evolutionary tree diagram **47**

see also hornworts; liverworts; mosses

bulbils 204

C

Callaghan, Des 209

calyptras, *Oncophorus integerrimus 26–27*

Canada, *Takakia lepidozioides* (moss, British Columbia) *52–53*

Canary Islands

mosses, *Pseudomalia webbiana 4–5*

capillary action 21, 126

carbon 83, 139, 142–143, 144

carbon sink 144

carbon dioxide 16, 22, 69, 79, 83, 143

Caribbean, *Pseudomarsupidium decipiens* (liverwort) *174* (175)

carnivorous plants 102, 106

carnivorous liverworts hypothesis 78, 106–107, *108–109*

carrion 104–105

Celtic rainforests 152, 167, *168–169*

Central America

liverworts

Adelanthus lindenbergianus 190–191

epiphyllous liverworts *129*

tropical cloud forests 123

Chile

liverworts

Gackstroemia 163

Lepicolea 163

temperate rainforests (Magellanic Region) 159

China

liverworts

Plagiochila 192

Plagiochila carringtonii 192

rainforests (Eastern part) 153

chloroplasts 79

Circumpolar Boreo-arctic montane 183

climate change prevention 82, 83, 144, 208–209

cloud forests *see* tropical cloud forests

clubmosses 43, 47

coneheads 91, 97

copper mosses, *Scopelophila cataractae* 198

Dál Riata kingdom 167, 170

digital photography 209

disturbed and urban environments

different places

man-made habitats 197–199

tops of urban walls 195–196

tree trunks 197

waste ground/concrete surfaces 196–197

hornworts, arable hornworts 204–205

liverworts

arable liverworts 204–205

Riccia cavernosa 204 (205)

mosses

arable mosses 204–205

Brachythecium rutabulum 196–197

Bryum argenteum 197

Bryum capillare 61 (60), *194* (195), 195

Buxbaumia aphylla ('bug moss') 199, 202, *202–203*

Ceratodon purpureus 196

Didymodon 196

Ditrichum plumbicola 198, *200–201*

Grimmia pulvinata 195

heavy-metal mosses 198–199

Homalothecium sericeum 195

Hypnum cupressiforme 197

Orthotrichum diaphanum 197

Scopelophila cataractae 198

Tortula muralis 195

Zygodon 197

DNA 23, 28, 30–31, 46–47, 50, 88, 150, 163

  evolutionary tree diagram **47**

dung, dung flies 104–105, *104*

E

elaters 65

endemic genera 163

enzymes 106, 107

epiphyllous liverworts 128, *129*

evolution

  ancient plants theories 44–45

  DNA evidence 46–47, 88

  evolutionary tree diagram **47**, 50

  fossils 45, 47, 48–49

  'lower' plants bias 167

F

Faroe Islands, *Plagiochila carringtonii* (liverwort) 191–192, *192*

ferns 7, 43, 46, 47

flies, dung flies 104–105, *104*

flood prevention 84

food chains 88

forests

  Celtic rainforests 152, 167, *168–169*

  laurisilva forest *166* (167)

  moss forests 90–91

  'mossy forest' 118

  see also temperate rainforests; tropical cloud forests

fossils 45, 47, 48–49

  in amber 49

fungi 88, 91, 94, 97, 143

G

gametes 33

gametophytes

  description and function 25, 28, 30, 31, 42, 51

  and evolution 45

  liverworts, *Aneura mirabilis* ('ghostwort') *72–73*

  mosses

    Andreaea 59

    Buxbaumia ('bug moss') 202

    Hookeria lucens 29 (28)

    Oncophorus integerrimus 26–27

    Polytrichastrum sphaerothecium 24–25

    Sphagnum 54

gemmae *20* (21), 32, *33*, 204

Georgia (Caucasus), *Pseudomalia webbiana* (moss) *4–5*

Germany, *Ditrichum plumbicola* (moss) 198, *200–201*

glaciers 179, 180, 186, 189

Gondwana 157, 159, 162, 163

H

habitats
habitat destruction 197–198, 208–209
microhabitats 90, 189
see also moss habitats

Haeckel, Ernst
Kunstformen der Natur, Muscinae 6 (7)

heaths 187, 208
oceanic montane heath 187, 189, 191–192

heavy-metal mosses 198–199

Himalayas
liverworts
Anastrophyllum alpinum 192
Plagiochila carringtonii 191–192, *192*

oceanic montane heath 191–192

Hodgetts, Nick, *Mosses and Liverworts* (Porley and Hodgetts) 9

hornworts
characteristics
diversity 12
form and ecology 16
pyrenoids and blue-green bacteria 79
species, number of 34, 40, 79
terminology issue 14–15, 34
comparison with mosses
differences 34, 40
similarities 34
evolution
DNA evidence 46–47
evolutionary tree diagram **47**, 50
specific plants
arable hornworts 204–205
Megaceros gracilis 40–41
see also sporophytes

horticulture 82, 88

hydrogen 142, 143

I

Iceland
mosses
Bryoxiphium norvegicum ('sword moss') *112–113*
Kiaeria starkei 181 (180)
Oncophorus integerrimus 26–27
Pseudobryum cinclidioides 17 (16)
Pterigynandrum filiforme 92–93
insects 91, 106, 146
spore dispersal by 102, 104–105, *104*

Ireland
liverworts
Adelanthus lindenbergianus 190–191
Plagiochila carringtonii 191, *192*
Pseudomarsupidium decipiens 174 (175)

mosses

    Paludella squarrosa
206–207

    peat 83

oceanic montane heath
189, 191

rainforests 153, *168–169*

J

Japan

    moss gardens 82

    rainforests 153

K

Kato, Kengo 150

Kimmerer, Robin Wall,
Gathering Moss 9

L

La Réunion

    liverworts

        *Gottschelia
schizopleura 77*

        *Lepidolejeunea
delessertii 130–131*

        *Plagiochila terebrans
134–135*

    mosses

        *Deslooveria usagara
135*

        *Orthostichella
versicolor 119*

        *Pelekium versicolor
86–87*

        *Phyllogonium fulgens
120–121*

        *Porotrichum
madagassum 135*

latitude 177

laurisilva forest *166* (167)

lead mosses, *Ditrichum
plumbicola* 198, *200–201*

lichens 42–43, 99, *116–117*,
182

liverworts

    books on 9

    by type

        arable liverworts
204–205

        epiphyllous liverworts
128, 129

        fan liverworts 136

        Haplomitrium
(Haplomitriopsida)
65, 66

        leafy 38, 65, **65**, 74,
74–75, 76, 77, 78,
80–81, 128

        oceanic montane
heath liverworts 187,
189, 191–192

        snowbed liverworts
179

        thalloid 32, 34, 38, 39,
40, 65, **65**

        thalloid - complex
65, 67 (66), 68 (69),
69, 70, 79, 138–139,
204–205

        thalloid - simple (most
with thickened
thalluses) **65**, 71, 71

        thalloid - simple (most
with translucent
thalluses) **65**, 71,
72–73

        Treubia
(Haplomitriopsida)
65, 66, 66

    carnivorous liverworts
hypothesis 78, 106–107,
108–109

    characteristics

        diversity 12

form and ecology 16

species, number of 34, 38

terminology issue 14–15, 34

vegetative reproduction 32–33

water, capturing and controlling 84

water and leafy liverworts 21

comparison with mosses

differences 34, 38, 40

similarities 34

evolution

DNA evidence 46–47

evolutionary tree diagram **47**, 50

fossils 45, 48

habitats

main characteristics 114–115

mountains and tundra 179, 180, 182

New Zealand 164

oceanic montane heath 187, 189

Scotland 8

temperate rainforests 152, 153, 159, 167, 170, 175

threats to 209

tropical cloud forests 118, 128

wetlands 140, 146

see also gametophytes; liverworts by type and name; sporophytes

lobules 78, 106, 107

lunar cycle, and *Sphagnum* mosses 147

M

Macaronesia

Pseudomarsupidium decipiens (liverwort) *174* (175)

tropical cloud forests 173

Madeira

mosses

Leucodon treleasei 62–63

Myurium hochstetteri 166 (167)

Pseudomalia webbiana 4–5

microhabitats 90, 189

Mironov, Victor 147

mites 91, 94, 97

oribatid ('armoured') mites (*Phauloppia lucorum*) 94, *94–95*

moon, and *Sphagnum* mosses 147

moss habitats

main characteristics 114–115

see also animals; disturbed and urban environments; mountains and tundra; temperate rainforests; tropical cloud forests; wetlands

mosses

books on 8–9

by type

Andreaea (Andreaeopsida) ('lantern mosses') **51**, 59–60, 59

aquatic mosses 148, 150

arable mosses 204–205

arctic and antarctic mosses 182–183, 186

fan mosses 135–136, 161, 175

giant mosses 124–126

heavy-metal mosses 198–199

liverwort-like mosses 128, 133

moss pillars 150, 151

mountain mosses 187, 189

Oedipodium (Oedipodiales) **51**

peristomate mosses (active type) **51**, 60, 61, 61 (60), 62–63, 64, 148

peristomate mosses ('pepper-pot' / haircap mosses) **51**, 61, 61 (60), 64

pleurocarpous mosses 8, 49, 132, *132* (133), 136, 195, 196–197

rheophytes 148, 150

snowbed mosses 178–179, 179, 209

Sphagnum (Sphagnopsida) 7, **51**, 54, 58, 59, 61, 88, 115, 147

Sphagnum peat bogs 54, 58, 83, 139, 142–144, 146–147

Takakia (Takakiopsida) **51**, 53

Tetraphis (Tetraphidales) **51**, 64

tree mosses 136, 156 (157), 161

characteristics

diversity 7, 34

form and ecology 16

popular misrepresentations 12

species, number of 34

terminology issue 14–15, 34

water, obtaining and transporting 21–23, 90–91

comparison with liverworts and hornworts

differences 34, 38, 40

similarities 34

conservation issues 208–209

different from

algae 42

clubmosses 43

lichens 42–43

Spanish moss 43

evolution

ancient plants theories 44–45

DNA evidence 46–47, 88

evolutionary tree diagram **47**, 50

fossils 45, 47, 48–49

'lower' plants bias 167

lifecycle

gametophytes and sporophytes 25, 28

sexual reproduction 28, 30–31, 33

vegetative reproduction 31–33

use of

climate change prevention 82, 83, 144, 208–209

holding/controlling water 82, 84–85

nutrient cycles and food chains 88

peat bogs 83

pillow stuffing and other historical uses 88

see also gametophytes; moss habitats; mosses by type and name; sporophytes

moss pillars 150, 151

mountains and tundra

characteristics 176–177, 180

   snowbeds *178–179*, 179, 209

liverworts

   Adelanthus lindenbergianus 190–191

   Anastrophyllum alpinum 192

   Herbertus borealis 192, *193* (192)

   Marsupella brevissima 183

   oceanic montane heath liverworts 187, 189, 191–192

   Plagiochila carringtonii 191–192, *192*

   Pleurozia purpurea 192

mosses

   arctic and antarctic mosses 115, 180, 182–183, 186

   Bryum weigelii 184–185

   Conostomum tetragonum 183

   Encalypta ciliata 188 (189)

   Kiaeria falcata 183

Kiaeria starkei 181 (180)

mountain mosses 187, 189

Polytrichastrum sexangulare 183

regions

   Antarctica 177, 180

   Arctic region 177, 180

   Scotland *176–177*, *177–178*, 180

mRNA transcripts 23

muirburn 208

Myanmar, fossils in amber 49, *49* (48)

N

Natural History Museum, London, UK 8

nesting material 102

   wren with moss *103* (102)

New Caledonia

   mosses

     Euptychium 122 (123)

     Spiridens spiridentoides 127 (126)

Mt. Aopinie tropical cloud forest *123*

New Zealand

   liverworts 164

     treubia 66

   mosses

     Catharomnion ciliatum 162, 163

     Cladomnion ericoides 162, 163

     Cryptopodium bartramioides 162, 163

     Mniodendron comatum 156 (157)

     Ptychomnion aciculare ('pipe cleaner moss') *158* (159)

   rainforests *85*, 153, 157, 161

nitrogen 79, 88, 106, 142, 196, 209

North America

   oceanic montane heath 191

   rainforests (Western part) 153, 175

   Scouleria (moss) 148

   snowbeds 179

Norway

oceanic montane heath 189, 191

rainforests 153

nutrient cycles 88

Oceania 126

oceanic montane heath 187, 189, 191–192

oxygen 22, 142

Pacific Islands, tropical cloud forests 123

paleoendemics 163

Pangea 49

Papua New Guinea, tropical cloud forests 123, 124

peat bogs 54, 58, 83, 139, 142–144, 146–147

peristomate mosses 51, 60, 61, 61 (60), 62–63, 64, 148

permafrost 182

phosphorus 88

photosynthesis

green plants 16

hornworts 79

liverworts 69, 146

mosses 22, 58, 125, 135–136, 202

pillow stuffing 88

pleurocarps 8, 49, 132, *132* (133), 136, 195, 196–197

pollution 209

Porley, Ron, *Mosses and Liverworts* (Porley and Hodgetts) 9

protists 100, 102, 107

protons 143

pseudopodium 54, 59

pyrenoids 79

rainforests

Celtic rainforests 152, 167, *168–169*

see also temperate rainforests

reproduction

sexual reproduction 28, 30–31, 33

vegetative reproduction 31–33

rheophytes 148, 150

rhizoidal tubers 204

rhizoids 140

Rhododendron ponticum 209

rotifers 91, 97, 100

genus *Nothocla 101*

Royal Botanic Garden Edinburgh, Scotland 8

saltmarshes 142

Scandinavia, snowbeds 179

Scotland

Cairngorm plateau *176–177*

snowbed *178–179*, 180, 209

corries 189

liverworts 8, 209

Adelanthus lindenbergianus 190–191

Anastrophyllum alpinum 192

Drepanolejeunea hamatifolia 172–173

Herbertus borealis 192, *193* (192)

Plagiochila carringtonii 192

Pleurozia purpurea 108–109

mosses 7–8

Buxbaumia aphylla ('bug moss') 199, 202, *202–203*

Grimmia decipiens 116–117

Grimmia montana 116–117

Kiaeria starkei 181 (180)

Myurium hochstetteri 166 (167)

for wound dressing 88

oak woodland 152

oceanic montane heath 189, 191–192

peat 83

rainforests (Western Scotland) 153, 167, 171

scrub oak 152

setas 51, 54, 59

sexual reproduction 28, 30–31, 33

Siberia, early moss fossils 48–49

snowbeds *178–179*, 179, 209

soil creation *86–87*, 88

soil protection 84, 88

South America

liverworts

Adelanthus lindenbergianus 190–191

Pseudomarsupidium decipiens 174 (175)

mosses

Leptobryum wilsonii 150

pleurocarps 8

Ptychomnion cygnisetum 159

rainforests 153, 157, 159, 161, 162

tropical cloud forests (Andes) 123

South East Asia

pleurocarps 8

tropical cloud forests 123, 124–126

Spanish moss 43

spiders 91, 97

sporophytes

description and function 25, 28, 30–31, 33

dispersal of spores by insects 102, 104–105, *104*

and evolution 44–45

hornworts 40

Megaceros gracilis 40–41

liverworts 38, 40, 45, 65, 69

Aneura mirabilis ('ghostwort') *72–73*

Conocephalum conicum 68 (69)

Plagiochasma rupestre 70

Sphaerocarpos (balloonwort) 204–205

mosses 51

Andreaea ('lantern mosses') 60

arable mosses 204

Bryum capillare 195

Buxbaumia ('bug moss') 202

Ceratodon purpureus 196

Exsertotheca intermedia 132 (133)

Grimmia pulvinata 195

Hennediella heimii 60

Hookeria lucens 29 (28)

Leptostomum 161

Oncophorus integerrimus 26–27

Orthotrichum diaphanum 197

peristomes 64

pleurocarps 132, 195

Polytrichastrum sphaerothecium 24–25

Sphagnum 54, 59, 61

Splachnum vasculosum 105

Takakia 53

Tayloria gunnii 105

Tortula muralis 195

sporopollenin 28

springtails 91, 94, 97

Dicyrtoma fusca 96–97

Star Trek: Discovery 99

supraorganismic system 147

T

tardigrades ('water bears') 91, *98*, 99

Tasmania

  mosses

Tayloria gunnii 105, *105*

Weymouthia mollis 159

rainforests 153, 161

temperate rainforests

  characteristics 84, 114–115, 152–153

    map of temperate and boreal rainforests 153

    Northern rainforests 167–175

    Southern rainforests 157–166

liverworts

  Drepanolejeunea hamatifolia 171, *172–173*

  Gackstroemia 163, 164

  Harpalejeunea molleri 171

  Jubula hutchinsiae 173

  Lepicolea 163, 164

  Plagiochila 171

  Pseudomarsupidium decipiens 171, *174* (175)

  Schistochila 164

  Schistochila appendiculata 164, 165 (164)

mosses

Alsia californica 175

Catagonium 161

Catharomnion ciliatum 162, 163

Cladomnion ericoides 162, 163

Cryptopodium bartramioides 162, 163

Dendroalsia abietina 175, *175*

Dicranoloma 161

fan mosses 136

Glyphothecium 161

Hylocomiadelphus 171

Hylocomium splendens 170 (171), 171

Hypnodendraceae 161

Hypopterygiaceae 161

Hypopterygium didictyon 157

Isothecium myosuroides ('mouse-tail moss') 171

Leptostomum 161

Loeskeobryum brevirostre 171

Mniodendron comatum 156 (157)

Myurium hochstetteri 166 (167)

Neckera 175

Ptychomnion 161, 171

Ptychomnion aciculare ('pipe cleaner moss') *158* (159)

Ptychomnion cygnisetum 159

Rhizogonium 161

Rhizogonium distichum 160 (161)

Rhytidiadelphus 171

Rhytidiadelphus loreus 171

tree mosses 136, *156* (157)

Ulota 154–155, 171

Weymouthia 161

Weymouthia mollis 159

regions

Australia, Eastern part and Tasmania 153, 157, 159, 161, 162

China, Eastern part 153

Ireland 153, *168–169*

Japan 153

New Zealand *85*, 153, 161

North America, Western part 153, 175

Norwegian Fiordland 153

Scotland, Western part 153, 167, 171

South America, Southern part 153, 157, 159, 161, 162

Wales *168–169*

totipotency 31–32, 186

tourism 209

tree lines 176–177, 187, 189, 192

tropical cloud forests

characteristics 84, 114–115, 118, 123

liverworts

epiphytes (epiphyllous liverworts) 128, *129*

fan liverworts 136

Lepidolejeunea delessertii 130–131

Plagiochila terebrans 134–135

Porellales 128

mosses

Dawsonia 124–126, *124, 125* (124)

Deslooveria usagara 135

Euptychium 122 (123)

fan mosses 135–136

giant mosses 124–126

Hookeriales 133

Hypopterygium tamarisci 137 (136)

liverwort-like mosses 128, 133

Meteoriaceae 126

Orthostichella versicolor 119

Phyllogonium fulgens 120–121

Porotrichum madagassum 135

Spiridens 126

Spiridens spiridentoides 127 (126)

tree mosses 136

Ulota fulva 154–155

regions and places

Africa, sub-Saharan mountains 123

Australia, Eastern Queensland 123

New Caledonia, Mt. Aopinie *123*

Pacific Islands 123

South American Andes 123

South East Asian islands 123, 124–126

tubers 32

tun state 99, 100

tundra *176–177*, 177, 180

see also mountains and tundra

U

UK (United Kingdom) *see* Britain

urban environments *see* disturbed and urban environments

V

vascular system 21, 125–126

vegetative reproduction 31–33

W

Wales
  cwms 189
  liverworts, *Trichocolea tomentella 76*
  mosses
    Ditrichum plumbicola 198, *200–201*
    Scopelophila cataractae 198

oceanic montane heath 189, 191

rainforests *168–169*

water
  and green plants 16
  and mosses
    holding and controlling 82, 84–85
    obtaining and transporting 21–23, 90–91
  and tardigrades 99
  see also wetlands

Wesołowski, Tomasz 102

west wind drift 157

wetlands
  different places
    fresh water 140
    by lakes, rivers, waterfalls, springs 140
    saltmarshes and under sea-spray 142
    Sphagnum peat bogs 34, 58, 83, 139, 142–144, 146–147

liverworts
  Aneura mirabilis ('ghostwort') *72–73*, 146
  Marchantia polymorpha subspecies

  montivagans 67 (66), *138–139*
  Odontoschisma sphagni 144, 146

mosses
  aquatic mosses 148, 150
  Calliergon cordifolium 138–139
  Fontinalis 148, *148*
  Fontinalis antipyretica 148, *149* (148)
  Leptobryum wilsonii 150
  moss pillars 150, *151*
  Schistidium maritimum 140–141
  Scouleria 148
  Sphagnum cuspidatum ('drowned kitten moss') 146
  Sphagnum skyense 145 (144)

Wierzcholska, Sylwia 102

worms 91, 94, 97

wound dressing 88

Z

zinc mosses 198

INDEX TO SCIENTIFIC NAMES OF MOSSES AND LIVERWORTS

A

Achrophyllum dentatum 20 (21)

Adelanthus lindenbergianus 190–191

Alsia californica 175

Anastrophyllum alpinum 192

Anastrophyllum hellerianum 33

Andreaeobryum macrosporum 60

Aneura mirabilis ('ghostwort') *72–73*, 146

Apotreubia 66

Atrichum undulatum 61 (60)

B

Biantheridion undulifolium 80–81

Blasia 79

Brachythecium rutabulum 196–197

Breutelia gnaphalea 36–37

Bryoxiphium norvegicum ('sword moss') *112–113*

Bryum argenteum 197

Bryum capillare 61 (60), *194* (195), 195

Bryum weigelii 184–185

Buxbaumia ('bug moss') 202

Buxbaumia aphylla 199, *202–203*

C

Calliergon cordifolium 138–139

Catagonium 161

Catharomnion ciliatum 162, 163

Ceratodon purpureus 196

Cladomnion ericoides 162, 163

Colura 106–107

Conocephalum conicum 68 (69)

Conoscyphus trapezioides 74–75

Conostomum tetragonum 183

Cryptopodium bartramioides 162, 163

Cyathodium cavernarum 39

D

Dawsonia 124–126, *124*, *125* (124)

Dendroalsia abietina 175, *175*

Deslooveria usagara 135

Dicranoloma 161

Didymodon 196

Ditrichum plumbicola 198, *200–201*

Drepanolejeunea hamatifolia 171, *172–173*

Encalypta ciliata 186

Euptychium 122 (123)

Exsertotheca intermedia 132 (133)

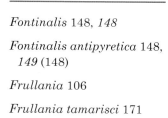

Fontinalis 148, *148*

Fontinalis antipyretica 148, *149* (148)

Frullania 106

Frullania tamarisci 171

G

Gackstroemia 163, 164

Glyphothecium 161

Gongylanthus ericetorum 35 (34)

Gottschelia schizopleura 77

Grimmia decipiens 116–117

Grimmia montana 116–117

Grimmia pulvinata 195

Harpalejeunea molleri 171

Hennediella heimii 60

Herbertus borealis 192, *193* (192)

Homalothecium sericeum 195

Hookeria lucens 29 (28)

Hookeriales 133

Hylocomiadelphus 171

Hylocomium splendens 170 (171), 171

Hypnodendraceae 161

Hypnodendrales *49* (48)

Hypnum 88

Hypnum cupressiforme 197

Hypnum revolutum 88

Hypopterygiaceae 161

Hypopterygium didictyon 157

Hypopterygium tamarisci 137 (136)

I

Isothecium myosuroides ('mouse-tail moss') 171

J

Jubula hutchinsiae 173

Jungermanniales 128

K

Kiaeria falcata 183

Kiaeria starkei 181 (180)

L

Lejeuneaceae 172–173

Lepicolea 163, 164

Lepidolejeunea delessertii 130–131

Leptobryum wilsonii 150

Leptostomum 161

Leucodon treleasei 62–63

Loeskeobryum brevirostre 171

M

Marchantia 32

Marchantia polymorpha subspecies *montivagans* 67 (66), *138–139*

Marsupella brevissima 183

Meteoriaceae 126

Mniodendron comatum 156 (157)

Moerckia flotoviana 71

Myurium hochstetteri 166 (167)

N

Neckera 175

O

Odontoschisma sphagni 144, 146

Oedipodium griffithianum 33

Oncophorus integerrimus 26–27

Orthostichella versicolor 119

Orthotrichum diaphanum 197

P

Paludella squarrosa 206–207

Pelekium versicolor 86–87

Petalophyllum ralfsii 13 (12)

Philonotis rigida 15

Phyllogonium fulgens 120–121

Plagiochasma rupestre 70

Plagiochila 171

Plagiochila carringtonii 191–192, 192

Plagiochila terebrans 134–135

Pleurozia 106–107

Pleurozia gigantea 108–109

Pleurozia purpurea 108– 109, 192

Polytrichaceae *24–25*, 125–126

Polytrichastrum sexangulare 183

Polytrichastrum sphaerothecium 24–25

Porellales 128

Porotrichum madagassum 135

Pseudobryum cinclidioides 17 (16)

Pseudomalia webbiana 4–5

Pseudomarsupidium decipiens 171, *174* (175)

Pterigynandrum filiforme
92–93

Ptychomitrium polyphyllum
110–111

Ptychomnion 161, 171

Ptychomnion aciculare
('pipe cleaner moss') *158*
(159)

Ptychomnion cygnisetum
159

R

Rhizogonium 161

Rhizogonium distichum 160
(161)

Rhytidiadelphus 171

Rhytidiadelphus loreus 171

Riccia cavernosa 204 (205)

Roaldia revoluta 89, (88)

S

Schistidium maritimum
140–141

Schistochila 164

Schistochila appendiculata
164, 165 (164)

Schistostega pennata is
18–19

Schlotheimia ferruginea
10–11 (11)

Scopelophila cataractae 198

Scouleria 148

Sphaerocarpos
(balloonwort) 204–205,
205

Sphagnum 54–55, 58–59

Sphagnum capillifolium
56–57

Sphagnum cuspidatum
('drowned kitten moss')
146

Sphagnum rubellum 80–81

Sphagnum skyense 145
(144)

Spiridens 126

Spiridens spiridentoides
127 (126)

Splachnaceae 104–105, *104*

Splachnum rubrum 104

Splachnum vasculosum
105, *105*

T

Takakia lepidozioides
52–53

Tayloria gunnii 105, *105*

Tortula muralis 195

Trichocolea tomentella 76

U

Ulota 154–155, 171

Ulota fulva 154–155

V

Vetiplanaxis pyrrhobryoides
49 (48)

W

Weymouthia 161

Weymouthia mollis 159

Z

Zygodon 197

Acknowledgements

This book is as much a collection of images supplemented by text as it is text supplemented by images. The vast majority of these are macro photographs by Des Callaghan. Des is a professional bryologist as well as a highly gifted photographer, and it is the (possibly unique) combination of these skills that enables him to create such stunning pictures of mosses and their relatives. Des is spectacularly generous in allowing others to use his photographs and I am immensely grateful to him for this, as well as for his help in making the best quality versions of his images available for this project.

Many thanks to all of the bryologists and other scientists whose research is mentioned (directly or indirectly) within this book. Their love of and dedication to the world of mosses has informed and enriched these pages. I'm particularly grateful to Liz Kungu and David Bell, fellow bryologists who read initial drafts of the text and offered constructive criticism. The wider bryological community at RBGE (David Long, David Chamberlain, Laura Forrest, Diego Sánchez-Ganfornina and Isuru Kariyawasam) has also been supportive, as have members of the British Bryological Society.

Not all of the images in the book were created by Des Callaghan and I'm grateful to everyone who has agreed to allow us to use their photographs.

Many people at the Royal Botanic Garden Edinburgh have been involved in the production of the book, but I would especially like to thank my editor, Sarah Worrall, and the designer, Caroline Muir. Thanks also to Paula Bushell (who initially encouraged me to write the book for RBGE), Andrew Lindsay, Becky Yahr (my officemate and lichenological counterpart at RBGE) and Chris Ellis, head of cryptogamic plants and fungi. Outside RBGE I'd like to thank Erica Schwarz for proofreading the text.

Finally, thanks to Monique for her sufferance of my bryological and non-bryological nerdiness in all of its multiple manifestations!